Name _____ Points and lines

Points, Line Segments, Rays, and Lines

words	point Q	line segment RS	ray TU	line XY
diagrams	Q•	R•⎯⎯•S	T•→•U↘	←•⎯•→ X Y
symbols	Q	$\overline{RS}$	$\overrightarrow{TU}$	$\overleftrightarrow{XY}$

Use words and symbols to describe each diagram.

1. A•   B•⎯⎯•C

   _____ _____ _____ _____
   _____ _____ _____ _____

2.   •J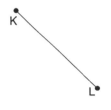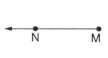

   _____ _____ _____ _____
   _____ _____ _____ _____

3. 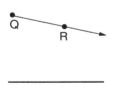      S•

   _____ _____ _____ _____
   _____ _____ _____ _____

Draw a picture for each symbol.

4. $\overline{WX}$ $\overrightarrow{YZ}$ $\overleftrightarrow{AB}$ C

Name _____ _____ Line Segments

Segment Lengths and Midpoints

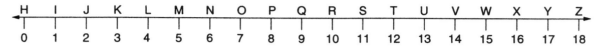

The endpoints of $\overline{LR}$ are 4 and 10.   The length of $\overline{LR}$ is 6.   The midpoint of $\overline{LR}$ is 7.

Use the number line.
Write the endpoints, length, and midpoint for each line segment.

	Segment	Endpoints	Length	Midpoint
1.	$\overline{KM}$	3, 5	2	4
2.	$\overline{OW}$			
3.	$\overline{RZ}$			
4.	$\overline{IS}$			
5.	$\overline{PT}$			
6.	$\overline{LZ}$			
7.	$\overline{MO}$			
8.	$\overline{JP}$			
9.	$\overline{TX}$			
10.	$\overline{NV}$			
11.	$\overline{HX}$			
12.	$\overline{HZ}$			

Name _____ Line segments

Properties of Line Segments

Congruence Property
Congruent segments have the same length.
$\overline{AB} = 3$ $\overline{CD} = 3$
$\overline{AB}$ is congruent to $\overline{CD}$. ($\overline{AB} \cong \overline{CD}$)

≅

Addition Property
$\overline{AB} = 3$ $\overline{BC} = 8$
$\overline{AB} + \overline{BC} = \overline{AC} = 11$

+

Look at each line segment. Use the line segment properties to complete the problems.

1. 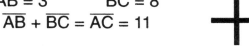 $\overline{EF} \cong$ ____ $\overline{EG} =$ ____ $\overline{FH} =$ ____

2.
```
  5    13     5
•——•————————•——•
I  J        K  L
```
$\overline{IK} =$ ____ $\overline{JL} =$ ____ $\overline{IJ} \cong$ ____

3.
```
  11    18     18    11
•——•———————•———————•——•
M  N       O       P  Q
```
$\overline{NO} \cong$ ____ $\overline{MO} =$ ____

$\overline{NP} =$ ____ $\overline{PQ} \cong$ ____ $\overline{OQ} =$ ____ $\overline{NQ} =$ ____ $\overline{MQ} =$ ____

4.
```
  12    24    12    24     18
•——•———————•———•———————•————•
R  S       T   U       V    W
```
$\overline{RS} \cong$ ____ $\overline{UW} =$ ____

$\overline{SU} =$ ____ $\overline{TV} =$ ____ $\overline{UV} \cong$ ____ $\overline{TW} =$ ____ $\overline{RV} =$ ____

5.
```
  4.9   7.1   6.3   7.1   4.9
•——•——————•—————•——————•——•
X  Y      Z     A      B  C
```
$\overline{XA} =$ ____ $\overline{ZB} =$ ____

$\overline{XY} \cong$ ____ $\overline{ZC} =$ ____ $\overline{AB} \cong$ ____ $\overline{AC} =$ ____ $\overline{YB} =$ ____

$\overline{AC} \cong$ ____ $\overline{YA} =$ ____ $\overline{XB} =$ ____ $\overline{YC} =$ ____ $\overline{XC} =$ ____

Name _____   Line relationships

Lots of Lines

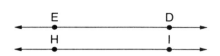

Parallel lines never meet.
They do not intersect.

$\overleftrightarrow{ED} \parallel \overleftrightarrow{HI}$ means line ED is parallel to line HI.

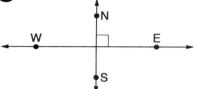

Perpendicular lines
intersect to form
right (90°) angles.

$\overleftrightarrow{NS} \perp \overleftrightarrow{WE}$ means line NS is perpendicular to line WE.

Study each diagram. Then write a symbol (∥ or ⊥) in each box. If the lines are not parallel or perpendicular, write an **X** in the box.

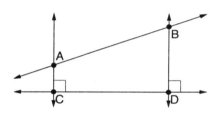

1. $\overleftrightarrow{AC}$ ☐ $\overleftrightarrow{CD}$ $\overleftrightarrow{AC}$ ☐ $\overleftrightarrow{BD}$
2. $\overleftrightarrow{DC}$ ☐ $\overleftrightarrow{BD}$ $\overleftrightarrow{AB}$ ☐ $\overleftrightarrow{BD}$

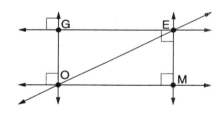

3. $\overleftrightarrow{GE}$ ☐ $\overleftrightarrow{EM}$ $\overleftrightarrow{GO}$ ☐ $\overleftrightarrow{EM}$
4. $\overleftrightarrow{OE}$ ☐ $\overleftrightarrow{OM}$ $\overleftrightarrow{MO}$ ☐ $\overleftrightarrow{EG}$
5. $\overleftrightarrow{GO}$ ☐ $\overleftrightarrow{EG}$ $\overleftrightarrow{EO}$ ☐ $\overleftrightarrow{GO}$

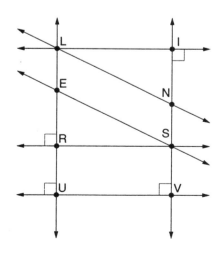

6. $\overleftrightarrow{LI}$ ☐ $\overleftrightarrow{UL}$ $\overleftrightarrow{LU}$ ☐ $\overleftrightarrow{RS}$
7. $\overleftrightarrow{IV}$ ☐ $\overleftrightarrow{SR}$ $\overleftrightarrow{UV}$ ☐ $\overleftrightarrow{IL}$
8. $\overleftrightarrow{SE}$ ☐ $\overleftrightarrow{NL}$ $\overleftrightarrow{NS}$ ☐ $\overleftrightarrow{UL}$
9. $\overleftrightarrow{ER}$ ☐ $\overleftrightarrow{RS}$ $\overleftrightarrow{IN}$ ☐ $\overleftrightarrow{SE}$

© Frank Schaffer Publications, Inc.   FS-10218 Introduction to Geometry

Mixed Practice With Lines

Complete.

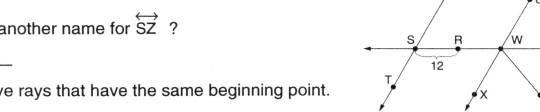

1. What is another name for $\overleftrightarrow{SZ}$?

   _____

2. Name five rays that have the same beginning point.

   _____ _____ _____ _____ _____

3. Name two pairs of intersecting lines.

   _____ and _____ ; _____ and _____

4. Name six line segments that lie on $\overleftrightarrow{SZ}$.

   _____ _____ _____ _____ _____ _____

5. R is the midpoint of $\overline{SW}$. What is the length of $\overline{SW}$? _____

6. W is the midpoint of $\overleftrightarrow{SZ}$. What is the length of $\overleftrightarrow{SZ}$? _____

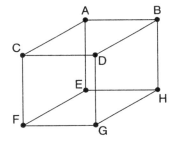

7. Name two line segments that are parallel to $\overline{AC}$.

   _____ _____

8. Name three line segments that include point F.

   _____ _____ _____

9. Name two line segments that are perpendicular to $\overline{BH}$. _____ _____

Use ⊥ or ∥ to complete.

10. $\overline{CD}$ ☐ $\overline{FG}$ $\overline{DG}$ ☐ $\overline{CD}$ $\overline{BH}$ ☐ $\overline{AB}$

11. $\overline{EH}$ ☐ $\overline{FG}$ $\overline{EA}$ ☐ $\overline{GD}$ $\overline{AC}$ ☐ $\overline{CD}$

Name _____ Naming angles

Angle ID

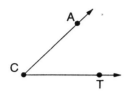 This angle may be named in 3 ways:
∠ACT
∠TCA
or ∠C

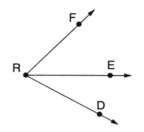 There are 3 angles shown here:
∠FRE or ∠ERF
∠ERD or ∠DRE
∠FRD or ∠DRF or ∠R

Write three names for each angle.

1.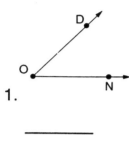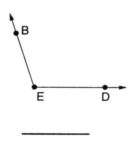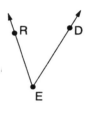

_____   _____   _____
_____   _____   _____
_____   _____   _____

Name three angles for each drawing.

2.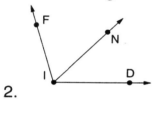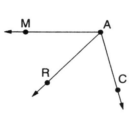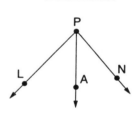

_____   _____   _____
_____   _____   _____
_____   _____   _____

Name six angles for each drawing.

3.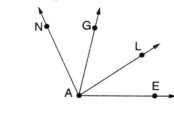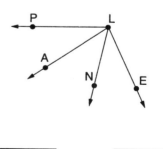

_____ _____   _____ _____
_____ _____   _____ _____
_____ _____   _____ _____

© Frank Schaffer Publications, Inc.   FS-10218 Introduction to Geometry

Name _____   Classifying angles

What's Your Angle?

Angles can be classified into 4 groups.
They are classified by their angle measures.

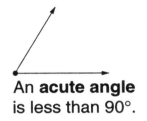   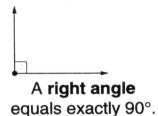   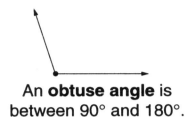   

An acute angle is less than 90°.   **A right angle** equals exactly 90°.   **An obtuse angle** is between 90° and 180°.   **A straight angle** equals exactly 180°.

Classify each angle as acute, right, obtuse, or straight.

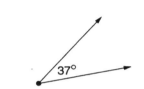   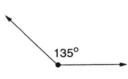   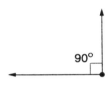

1. _____   _____   _____   _____

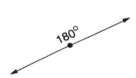   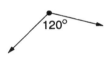   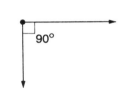

2. _____   _____   _____   _____

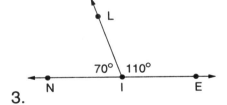   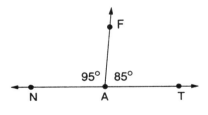

3.

∠LIN _____   ∠BOX _____   ∠FAN _____

∠LIE _____   ∠BOY _____   ∠FAT _____

∠NIE _____   ∠XOY _____   ∠NAT _____

Name _____ Angles

Properties of Angles

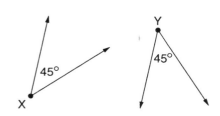

Congruence Property
Angles with the same measure (m) are congruent.
m∠X = 45°   m∠Y = 45°
∠X ≅ ∠Y

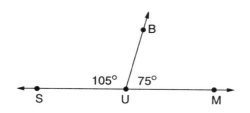

Addition Property
m∠BUM = 75°   m∠BUS = 105°
m∠BUM + m∠BUS = 180°
m∠BUM + m∠BUS = m∠SUM
75° + 105° = 180°

Use the angle properties to solve.

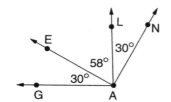

1. m∠NAE = _____   m∠LAG = _____
2. ∠NAL ≅ ∠ _____   m∠NAG = _____

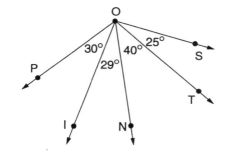

3. m∠PON = _____   m∠SON = _____
4. m∠TOI = _____   m∠TOP = _____
5. m∠SOP = _____   m∠SOI = _____

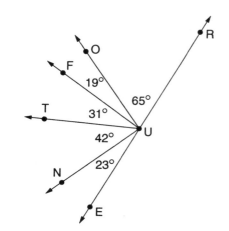

6. m∠FUR = _____   m∠OUT = _____
7. m∠TUE = _____   m∠EUR = _____
8. ∠TUE ≅ ∠ _____   m∠RUN = _____
9. m∠NUF = _____   m∠FUE = _____

© Frank Schaffer Publications, Inc.   FS-10218 Introduction to Geometry

Name _____   Angles

Complementary Angles

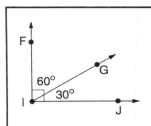

Complementary angles have a combined measure of 90°.
m∠FIG + m∠GIJ = m∠FIJ
60° + 30° = 90°
∠FIG and ∠GIJ are complementary angles.

Name the complementary angle for each angle.

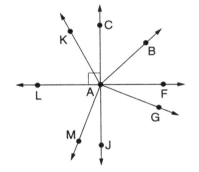

1. ∠CAB _____   ∠LAK _____

2. ∠LAM _____   ∠GAF _____

3. ∠FAB _____   ∠JAM _____

4. ∠CAK _____

Find the measure of the complement for each angle. Write an equation to help you.

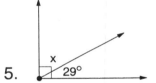 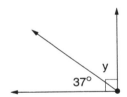 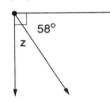

5.

$$x + 29° = 90°$$
$$x = 61°$$

_____   _____

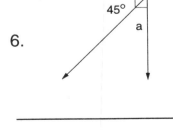 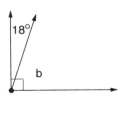 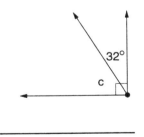

6.

_____   _____   _____

_____   _____   _____

© Frank Schaffer Publications, Inc.   9   FS-10218 Introduction to Geometry

Name _____   Angles

Supplementary Angles

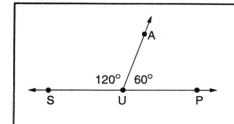

Supplementary angles have a combined measure of 180°.
m∠SUA + m∠AUP = m∠SUP
120° + 60° = 180°
∠SUA and ∠AUP are supplementary angles.

Draw lines between the letters to match each angle with its supplement.

1. 127°   35°   82°   94°   ⌐
    a   b   c   d   e

    f   g   h   i   j
    98°   53°   ⌐   145°   86°

Give the measure of each angle's supplement. Write an equation to help you.

2. 68° a   118° b   c 17°

   _____   _____   _____
   _____   _____   _____

3. d 79°   37° c   f 88°

   _____   _____   _____
   _____   _____   _____

© Frank Schaffer Publications, Inc.   FS-10218 Introduction to Geometry

Name _____ Angles

Vertical Angles

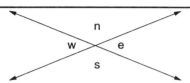

Vertical angles are congruent.   ∠n ≅ ∠s   ∠w ≅ ∠e
If m∠n = 140°, then m∠s = 140°.
Supplementary angles measure a total of 180°.
If m∠n = 140°, then m∠w = 40°.
If ∠w ≅ ∠e, then m∠e = 40°.

Find the measures of the missing angles.

1. 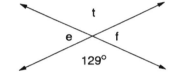   m∠t = __129°__   m∠e = __51°__   m∠f = __51°__

2. 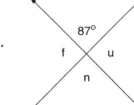   m∠f = _____   m∠u = _____   m∠n = _____

3. 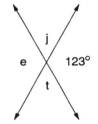   m∠j = _____   m∠e = _____   m∠t = _____

4. 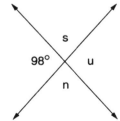   m∠a = _____   m∠c = _____   m∠t = _____

5. 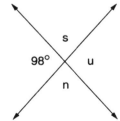   m∠s = _____   m∠u = _____   m∠n = _____

6. 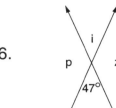   m∠i = _____   m∠p = _____   m∠z = _____

7. 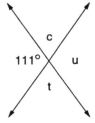   m∠c = _____   m∠u = _____   m∠t = _____

8. 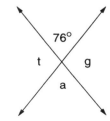   m∠t = _____   m∠a = _____   m∠g = _____

© Frank Schaffer Publications, Inc.   FS-10218 Introduction to Geometry

Name _____ Angles

Corresponding Angles

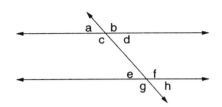

When a line intersects parallel lines, **corresponding angles** are congruent.

In the diagram at the left, the following pairs of angles are **corresponding**:
∠a and ∠e   ∠b and ∠f
∠c and ∠g   ∠d and ∠h
If m∠a = 70°, then m∠e = 70°.

Find the measures of the angles.

1.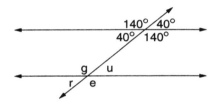

m∠g = _____   m∠u = _____
m∠e = _____   m∠r = _____

2.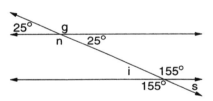

m∠n = _____   m∠g = _____
m∠i = _____   m∠s = _____

3.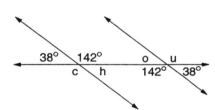

m∠u = _____   m∠o = _____
m∠c = _____   m∠h = _____

4.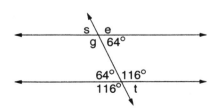

m∠s = _____   m∠e = _____
m∠g = _____   m∠t = _____

Name _____ Angles

Finding Angle Measures

Find the measures of the angles.
Use what you know about complementary, supplementary, and vertical angles to help you.

1.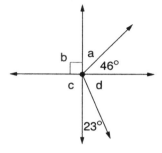

   m∠a = _____   m∠b = _____
   m∠c = _____   m∠d = _____

2.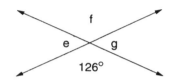

   m∠e = _____   m∠f = _____   m∠g = _____

3.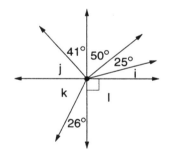

   m∠i = _____   m∠j = _____
   m∠k = _____   m∠l = _____

4.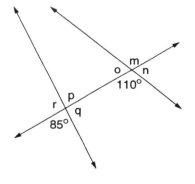

   m∠m = _____   m∠n = _____
   m∠o = _____   m∠p = _____
   m∠q = _____   m∠r = _____

© Frank Schaffer Publications, Inc.   FS-10218 Introduction to Geometry

Name _____ Angles

More Angle Measures

Find the measure of each angle.
Use what you know about supplementary, vertical, and corresponding angles to help you.

1.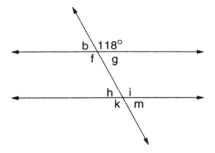

m∠b = _____   m∠f = _____
m∠g = _____   m∠h = _____
m∠i = _____   m∠k = _____
m∠m = _____

2.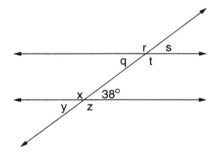

m∠r = _____   m∠s = _____
m∠q = _____   m∠t = _____
m∠x = _____   m∠y = _____
m∠z = _____

3.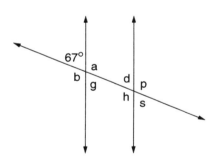

m∠a = _____   m∠b = _____
m∠d = _____   m∠h = _____
m∠p = _____   m∠g = _____
m∠s = _____

4.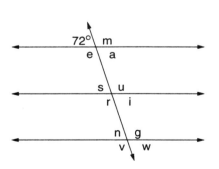

m∠m = _____   m∠e = _____
m∠a = _____   m∠s = _____
m∠u = _____   m∠r = _____
m∠i = _____   m∠n = _____
m∠g = _____   m∠v = _____
m∠w = _____

Name _____   Triangles

Classifying Triangles by Their Sides

Triangles can be classified by the lengths of their sides.

scalene
no equal sides

isosceles
2 equal sides

Little lines called hash marks show sides that are congruent (equal).

equilateral
3 equal sides

Write **scalene, isosceles,** or **equilateral** to classify each triangle. Draw hash marks (\) to show congruent sides.

A.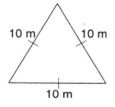

_____   _____   _____

B.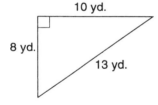

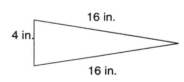

_____   _____   _____

C.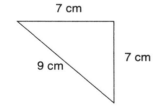

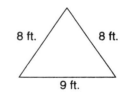

_____   _____   _____

In the space below, draw and label an equilateral triangle, an isosceles triangle, and a scalene triangle. Make hash marks to indicate congruent sides.

Name _____ Triangles

Classifying Triangles by Their Angles

Triangles can be classified by their angles.

acute triangle
all angles are acute

equiangular triangle
all angles are acute and congruent

obtuse triangle
one angle is an obtuse angle

right triangle
one angle is a right angle

Write **acute, equiangular, obtuse,** or **right** to describe each triangle.

A. 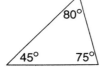   (60°, 90°, 30°)

_____   _____   _____

B. 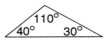   (60°, 60°, 60°)

_____   _____   _____

C. 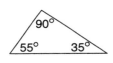   (82°, 82°, 16°)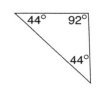

_____   _____   _____

Draw three different types of triangles. Label each as acute, equiangular, right, or obtuse.

D.

_____   _____   _____

Name _____ Triangles

Triangles by Graphic Design

Plot the ordered pairs and connect the points to construct each triangle.
Then classify each triangle by its sides and angles.

A.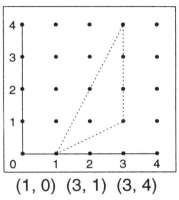
(1, 0) (3, 1) (3, 4)

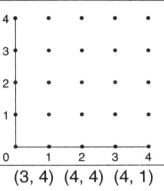
(3, 4) (4, 4) (4, 1)

Classify by the Sides
  scalene
  isosceles
  equilateral

Classify by the Angles
  acute
  equiangular
  obtuse
  right

_____

_____

B.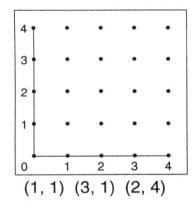
(1, 1) (3, 1) (2, 4)

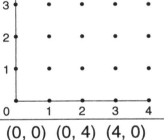
(0, 0) (0, 4) (4, 0)

(1, 1) (3, 1) (0, 4)

_____

_____

_____

C.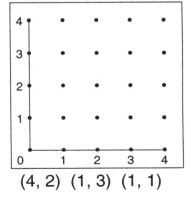
(4, 2) (1, 3) (1, 1)

(4, 4) (2, 1) (1, 3)

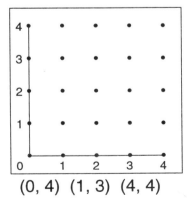
(0, 4) (1, 3) (4, 4)

_____

_____

_____

© Frank Schaffer Publications, Inc.   FS-10218 Introduction to Geometry

Name _____ Triangles

"Tri" These Missing Angles

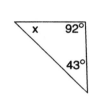

Every triangle has a total angle measure of 180°.
To find the measure of an unknown angle, add the two known angle measures and subtract their sum from 180°.
92° + 43° = 135°
180° − 135° = 45°
x = 45°

Find the measure of each missing angle.

A.

42°, 90°, y

60°, z, 60°

78°, 78°, a

B.

37°, b, 93°

c, 107°, 39°

73°, 59°, d

C.

31°, e, 15°

f, 49°, 69°

g, 98°, 41°

© Frank Schaffer Publications, Inc. FS-10218 Introduction to Geometry

Name _____ Triangles

Interior and Exterior Angles

In the diagram below, angles a, b, and c are **interior angles**. Angle x is an **exterior angle**. The sum of the measures of the interior angles a and b is equal to the measure of the exterior angle x.

$m\angle a + m\angle b = m\angle x$
Let $m\angle a = 89°$ and $m\angle b = 37°$.
$89° + 37° = m\angle x = 126°$
Since $\angle x$ and $\angle c$ are supplementary angles,
$m\angle c = 180° - 126° = 54°$

Find the missing angle measures. Use what you know about interior, exterior, and supplementary angles.

1.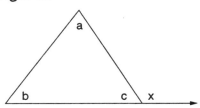

   m∠s _____   m∠r _____

2.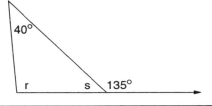

   m∠g _____   m∠h _____

3.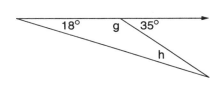

   m∠t _____   m∠r _____

4.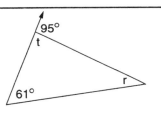

   m∠m _____   m∠n _____

5.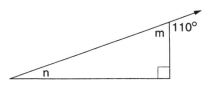

   m∠k _____   m∠m _____

© Frank Schaffer Publications, Inc.   FS-10218 Introduction to Geometry

Name _____ Congruent triangles

Proving Congruence by Side, Side, Side

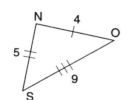 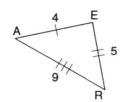

Triangles are **congruent** if their corresponding sides are congruent.
$\overline{NO} \cong \overline{EA}$
$\overline{OS} \cong \overline{AR}$
$\overline{SN} \cong \overline{RE}$
$\triangle NOS \cong \triangle EAR$

Use the **Side, Side, Side Rule (SSS)** to prove that the triangles in each pair are congruent. Draw hash marks to show the congruent sides. Then write a congruence statement for each pair of triangles.

1.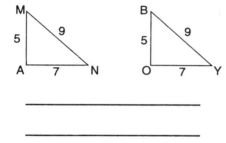

2.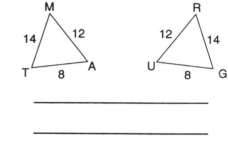

3.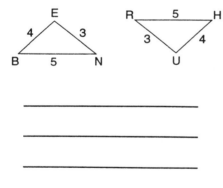

4.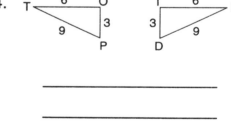

5.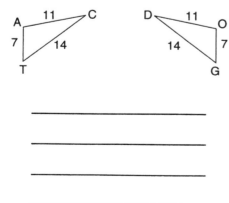

6.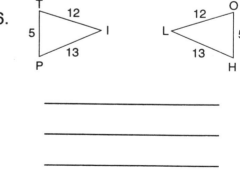

© Frank Schaffer Publications, Inc.   20   FS-10218 Introduction to Geometry

Name _____ Congruent triangles

Proving Congruence by Side, Angle, Side

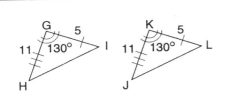

Two triangles are congruent if two sides and the angle they form on one triangle are congruent to two sides and the angle they form on the other triangle.

$\overline{GH} \cong \overline{KJ}$
$\overline{GI} \cong \overline{KL}$
$\angle G \cong \angle K$
$\triangle GHI \cong \triangle JKL$

Use the **Side, Angle, Side Rule (SAS)** to prove that the triangles in each pair are congruent. Draw hash marks to show the congruent sides and angles. Then write the congruence statements for each pair of triangles.

1.

2.

3.

4.

5.

6.

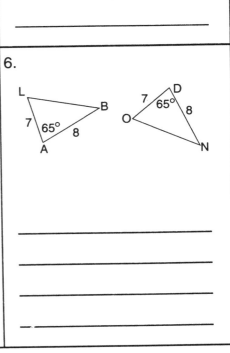

© Frank Schaffer Publications, Inc.   21   FS-10218 Introduction to Geometry

Proving Congruence by Two Angles and a Side

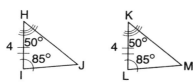

Angle, Side, Angle (ASA)
Triangles are congruent if two angles of one triangle and the side between the angles are congruent to two angles of the other triangle and the side between the angles.
ΔHIJ ≅ ΔKLM

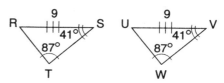

Angle, Angle, Side (AAS)
Triangles are congruent if two angles of one triangle and a side that is not between the angles are congruent to two angles of the other triangle and a side that is not between the angles.
ΔRST ≅ ΔUVW

Use **ASA** or **AAS** to prove that the triangles in each pair are congruent. Draw hash marks to show the congruent sides and angles. Then write a congruence statement for each pair of triangles and write ASA or AAS. If you cannot prove congruence, write **NOT**.

1.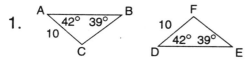

_____

2.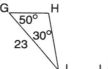

_____

3.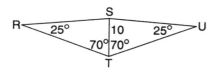

_____

4.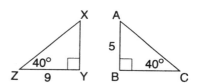

_____

5.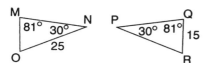

_____

6.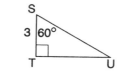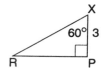

_____

Name _____ Congruent triangles

Are They Congruent?

Are the triangle pairs congruent? If so, write **SSS, SAS, ASA,** or **AAS.** If you cannot prove congruence, write **NOT.**

1. 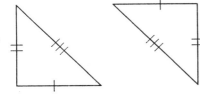 SSS

2. 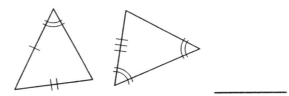 _____

3. 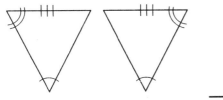 _____

4. 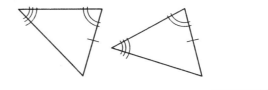 _____

5. 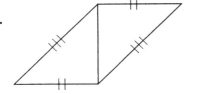 _____

6. 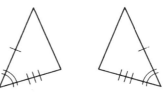 _____

7. 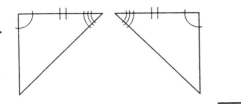 _____

8. 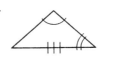 _____

9. 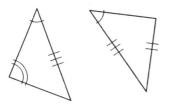 _____

10. 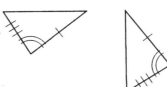 _____

© Frank Schaffer Publications, Inc. FS-10218 Introduction to Geometry

Name _____ Congruent triangles

Proving Congruence by Hypotenuse and Leg

Right triangles are congruent if the hypotenuse and a leg of one right triangle are congruent to the hypotenuse and a leg of another right triangle.
$\overline{FE} \cong \overline{IH}$ and $\overline{DE} \cong \overline{GI}$, so
$\triangle DEF \cong \triangle GIH$

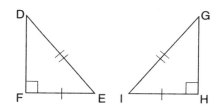

Use the **Hypotenuse and Leg Rule** to prove congruence. If the triangles are congruent, write **HL** on the line. If you cannot prove congruence, write **Not.**

1.
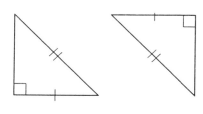

_____

2.
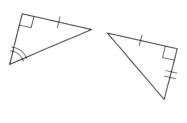

_____

3.
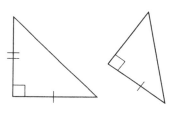

_____

4.
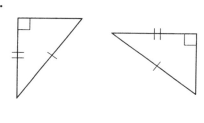

_____

5.
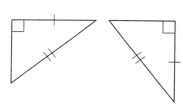

_____

6.
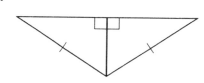

_____

7.
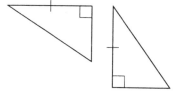

_____

8.
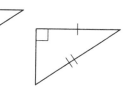

_____

© Frank Schaffer Publications, Inc.   FS-10218 Introduction to Geometry

Name _____ Congruent triangles

Name the Congruence Rule!

Are the triangles in each pair congruent? If so, write **SSS, SAS, ASA, AAS,** or **HL** to show how you proved congruence. If you cannot prove congruence, write **NOT**.

1.

   _____

2.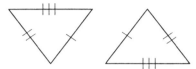

   _____

3.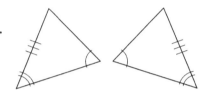

   _____

4.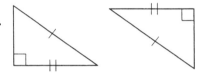

   _____

5.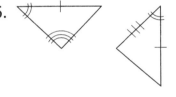

   _____

6.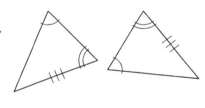

   _____

7.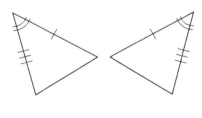

   _____

8.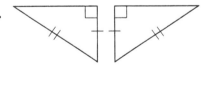

   _____

9.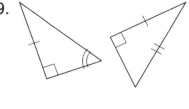

   _____

10.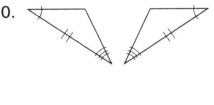

    _____

Name _____   Polygons

Quadrilaterals

A **polygon** is a closed plane figure formed by three or more line segments. Any four-sided polygon is a **quadrilateral**. Study the quadrilaterals in the box below.

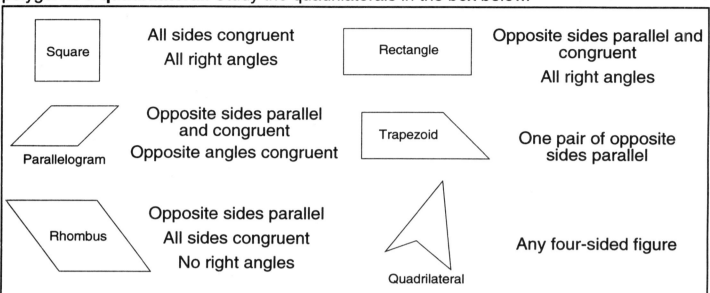

Use the figure at the right to complete items 1–4.

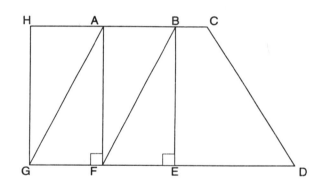

1. Name a square. _____

2. Name two rectangles.

   _____   _____

3. Name a parallelogram. _____

4. Name six trapezoids.

   _____ _____ _____ _____ _____ _____

Draw one line segment in each figure below to form the figures listed.

5.

   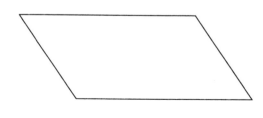   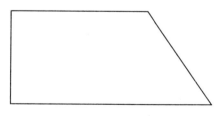

two trapezoids   two rhombuses   one square and
                                                  one trapezoid

© Frank Schaffer Publications, Inc.   26   FS-10218 Introduction to Geometry

Name _____   Polygons

Quadrilaterals by Graphic Design

Plot the ordered pairs and connect the points to construct each quadrilateral. Then write **rectangle, square, parallelogram, trapezoid,** or **rhombus** to identify each quadrilateral.

A.

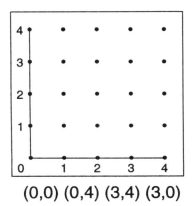

(0,0) (0,4) (3,4) (3,0)

(1,1) (1,4) (3,3) (3,0)

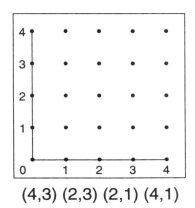

(4,3) (2,3) (2,1) (4,1)

B.

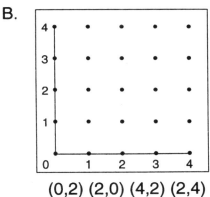

(0,2) (2,0) (4,2) (2,4)

(0,0) (2,4) (3,1) (4,3)

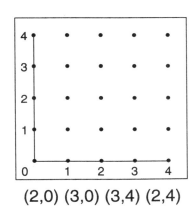

(2,0) (3,0) (3,4) (2,4)

C.

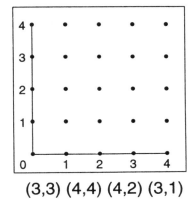

(3,3) (4,4) (4,2) (3,1)

(0,4) (1,0) (3,0) (4,4)

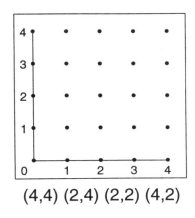

(4,4) (2,4) (2,2) (4,2)

© Frank Schaffer Publications, Inc.   27   FS-10218 Introduction to Geometry

Name _____ Polygons

Polygon Construction

Draw lines between the dots to show a polygon that fits each description. If you have a geoboard, you may use it to help you.

A. 3 sides
   1 right angle

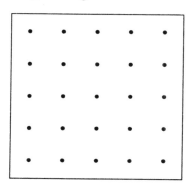

B. 6 sides
   2 equal sides

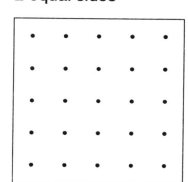

C. 4 sides
   4 right angles

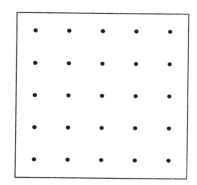

D. 5 unequal sides

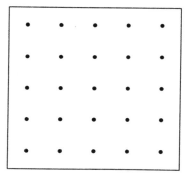

E. 8 sides

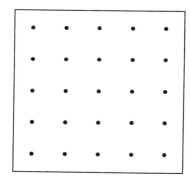

F. 3 sides
   no right angles

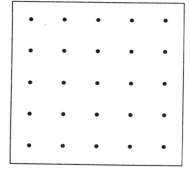

Draw your own polygons. Write their descriptions.

G. _____   _____   _____
   _____   _____   _____

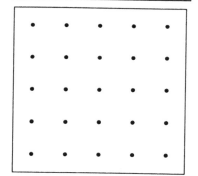

© Frank Schaffer Publications, Inc.   FS-10218 Introduction to Geometry

Name _____ Polygons

More Polygon Construction

Draw lines between the dots to show a polygon that fits each description.
If you have a geoboard, you may use it to help you.

A. 3 sides
   2 sides perpendicular

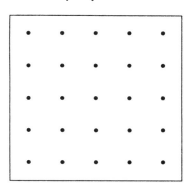

B. 4 sides
   2 parallel sides

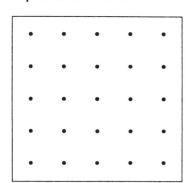

C. 4 sides
   2 pairs of parallel sides

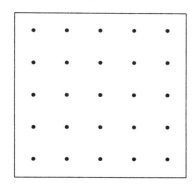

D. 6 sides
   3 pairs of parallel sides

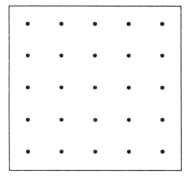

E. 5 sides
   2 parallel sides

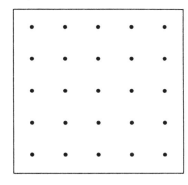

F. 4 sides
   2 sides perpendicular

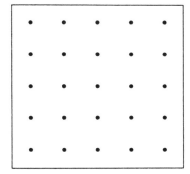

Draw your own polygons. Write their descriptions.

G. _____   _____   _____
   _____   _____   _____

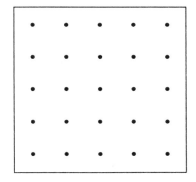 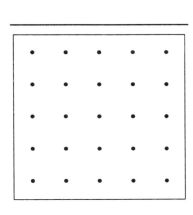

© Frank Schaffer Publications, Inc.   29   FS-10218 Introduction to Geometry

Name _____ Polygons

Sides, Vertices, and Diagonals of Polygons

A **vertex** is a point where the sides of a polygon meet.
A **diagonal** is a line segment that joins two nonadjacent vertices.

A square has 4 sides and 4 vertices. From any one vertex of a square, it is possible to draw only 1 diagonal. The diagonal will form 2 triangles. Since every triangle has a total of 180°, a square has a total of 360°.

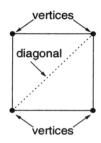

Study each figure below. Then complete the table.

Figure	Number of sides	Number of vertices	Diagonals from 1 vertex	Number of triangles formed	Total number of degrees
A.					
B.					
C.					
D.					
E.					
F.					

© Frank Schaffer Publications, Inc. 30 FS-10218 Introduction to Geometry

Name _____ Polygons

Name That Polygon!

Polygons are classified by the number of sides they have and by the lengths of their sides.

> 3 sides—triangle 7 sides—heptagon
> 4 sides—quadrilateral 8 sides—octagon
> 5 sides—pentagon 9 sides—nonagon
> 6 sides—hexagon 10 sides—decagon
> A polygon with all its sides of equal length
> is called a **regular** polygon.

Classify each polygon. If it is a regular polygon, write **R** on the drawing.

A.
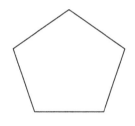 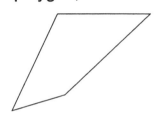 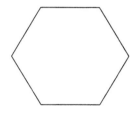

_____ _____ _____

B.
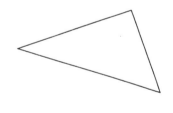 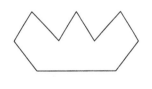 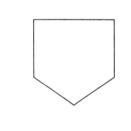

_____ _____ _____

C.
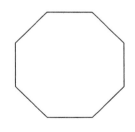  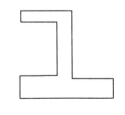

_____ _____ _____

On the back of this page, draw a hexagon, an octagon, and a regular quadrilateral.

Name _____ Polygons

Polygon Riddles

Draw and label the solution to each riddle.

A. I am a quadrilateral with four equal sides and no right angles. What am I?

_____

B. I am a six-sided polygon with sides of assorted lengths. What am I?

_____

C. I am an eight-sided polygon with all sides congruent. What am I?

_____

D. I am a quadrilateral with four right angles and two pairs of congruent and parallel lines. What am I?

_____

E. I have 10 sides that are of different lengths. What am I?

_____

F. I am a quadrilateral with no right angles and two pairs of congruent and parallel lines. What am I?

_____

G. I am a quadrilateral with one pair of parallel sides. What am I?

_____

H. I have two sides more than a triangle. All sides are the same length. What am I?

_____

© Frank Schaffer Publications, Inc.   FS-10218 Introduction to Geometry

Name _____   Perimeter

Dot-to-Dot Perimeter

Perimeter is the distance around a polygon. Draw a polygon with each of the given perimeters. On this page, each unit equals the distance between two dots on the grid.

A.
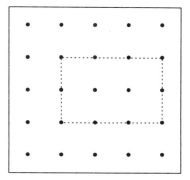
Perimeter = 10 units

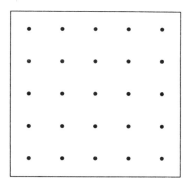
Perimeter = 12 units

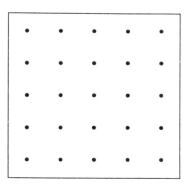
Perimeter = 16 units

B.
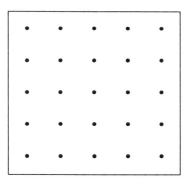
Perimeter = 8 units

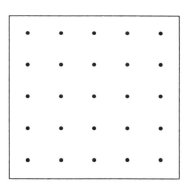
Perimeter = 4 units

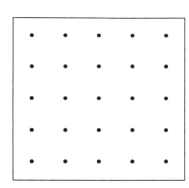
Perimeter = 15 units

C.
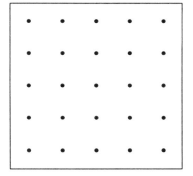
Perimeter = 6 units

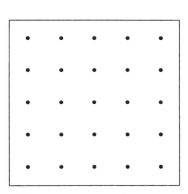
Perimeter = 18 units

Perimeter = 14 units

© Frank Schaffer Publications, Inc.   FS-10218 Introduction to Geometry

Name _____   Polygons

Perimeter of Regular Polygons

Remember! All sides of a regular polygon are congruent.

To find the perimeter (P) of a regular polygon, multiply the length of one side by the number of sides.

7 cm × 6 sides = 42 cm

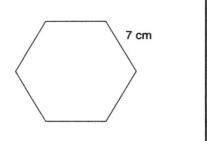

A.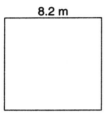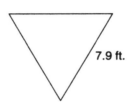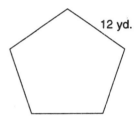

P = _____   P = _____   P = _____

B.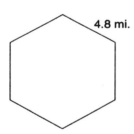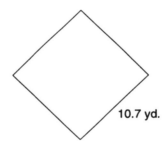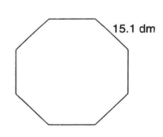

P = _____   P = _____   P = _____

C.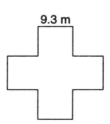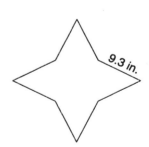

P = _____   P = _____   P = _____

Name _____ Perimeter

Polygon Perimeters

To find the perimeter (P) of a polygon that is not regular, add the lengths of its sides.

A.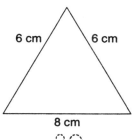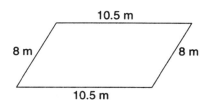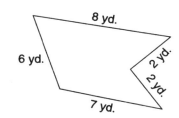

P = __20 cm__   P = _____   P = _____

B.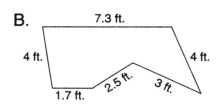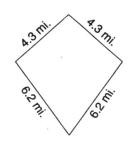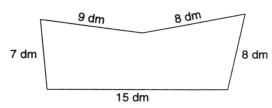

P = _____   P = _____   P = _____

C.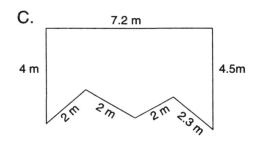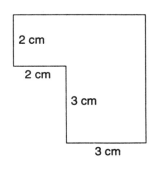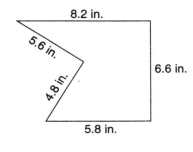

P = _____   P = _____   P = _____

D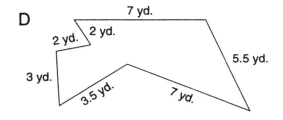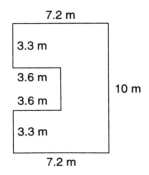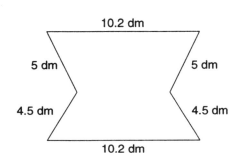

P = _____   P = _____   P = _____

Name _____ Circumference

Circumference of a Circle

Circumference (C) is the distance around a circle. There are two formulas you can use to find circumference. Each formula uses π. Let π = 3.14.

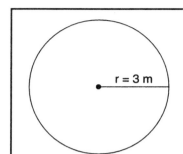

C = π • 2 • r
C = 3.14 • 2 • 3
C = 18.84 m

C = π • d
C = 3.14 • 7.5
C = 23.55 mm

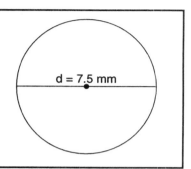

Find the circumference of each circle. Round to the nearest tenth. You may use a calculator.

1.
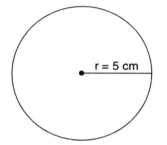 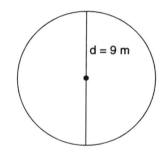 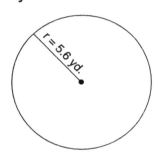

C = _____   C = _____   C = _____

2.
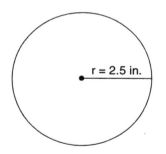  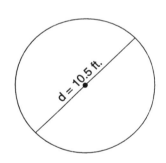

C = _____   C = _____   C = _____

3.
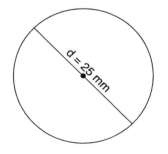 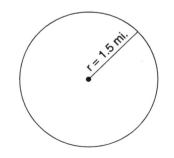 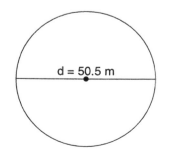

C = _____   C = _____   C = _____

© Frank Schaffer Publications, Inc. FS-10218 Introduction to Geometry

Name _____ Area

Counting Square Units

To find the area of a figure, count the square units in it. Write the answers in square units.

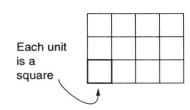

Each unit is a square

The area is 12 square units.
Write $A = 12\ u^2$ ← squared units

Count square units to find the area of each figure.

1.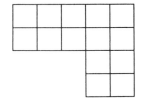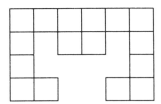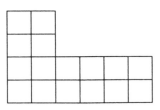

   A = _____   A = _____   A = _____

2.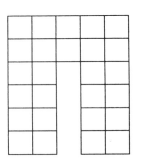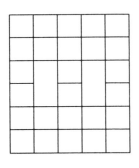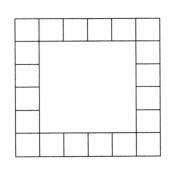

   A = _____   A = _____   A = _____

3.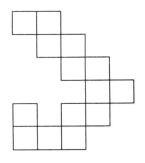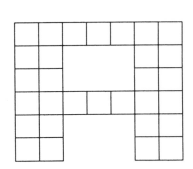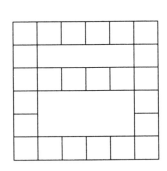

   A = _____   A = _____   A = _____

Name _____ Area

Find the Area!

Find the area by counting square units (squares). You may use a geoboard or draw lines to help you. Write your answers in square units.

A.
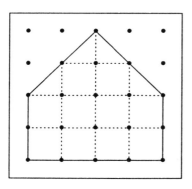
A = ___12 u²___

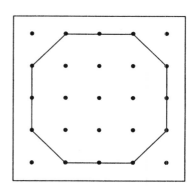
A = _____

A = _____

B.
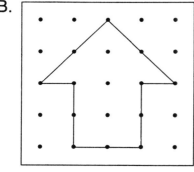
A = _____

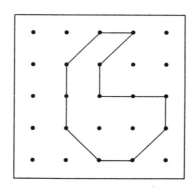
A = _____

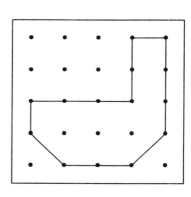
A = _____

C.
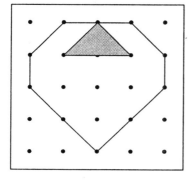
A = _____

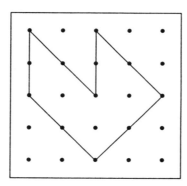
A = _____

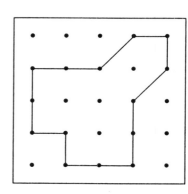
A = _____

© Frank Schaffer Publications, Inc.   FS-10218 Introduction to Geometry

Name _____ Area

Area of Rectangles and Squares

To find the **area** (A) of any rectangle, multiply its length times its width. Remember that a square is a rectangle.

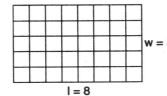

A = length • width
A = 8 • 5
A = 40 u²

Find the area of each square or rectangle. Write your answers in square units.

A.
6 in.
15 in.
A = ___90 in.²___

B.
12.2 yd.
8 yd.
A = _____

C.
7.4 m
A = _____

D.
6.2 ft.
8.7 ft.
A = _____

E.
8.9 m
A = _____

F.
3.8 dm
12.6 dm
A = _____

G.
12 cm
17 cm
A = _____

H.
10.6 mm
7.5 mm
A = _____

© Frank Schaffer Publications, Inc.   FS-10218 Introduction to Geometry

Name _____ Area

Area of Combined Shapes

Sometimes you have to separate a figure into smaller figures before you can find the area. Find the area of the smaller figures and add their areas to find the total area.

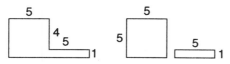

$A = (5 \cdot 5) + (5 \cdot 1)$
$A = 25 + 5$
$A = 30\ u^2$

Find the total area of each figure. First draw lines to show the smaller figures. Then write an equation for finding the area. Give the area in square units (u^2).

A.

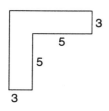   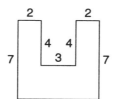   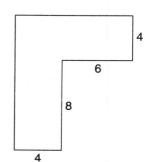

Area = _____   Area = _____   Area = _____

B.

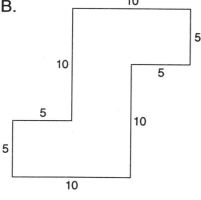   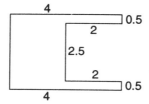   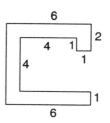

Area = _____   Area = _____   Area = _____

© Frank Schaffer Publications, Inc.   40   FS-10218 Introduction to Geometry

Name _____   Area

Area of Parallelograms

To find the area of a parallelogram, multiply its base times its height.

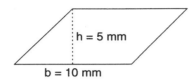

$A = b \cdot h$
$A = 10 \text{ mm} \cdot 5 \text{ mm}$
$A = 50 \text{ mm}^2$

Find the area of each parallelogram. Write your answers in square units.

A.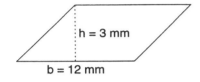

A = _____

B.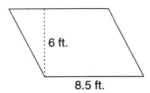

A = _____

C.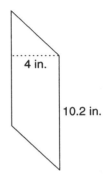

A = _____

D.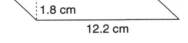

A = _____

E.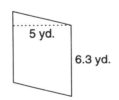

A = _____

F.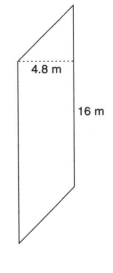

A = _____

G.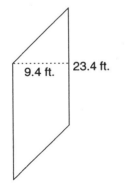

A = _____

H.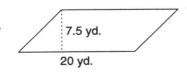

A = _____

© Frank Schaffer Publications, Inc.   41   FS-10218 Introduction to Geometry

Name _____ Area

Area of Triangles

To find the area of a triangle, multiply $\frac{1}{2}$ its base times its height.

$A = \frac{1}{2} \cdot b \cdot h$

$A = \frac{1}{2} \cdot 10 \cdot 3$

$A = \frac{1}{2} \cdot 30$

$A = 15$ cm²

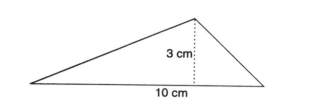

Find each area using the formula $A = \frac{1}{2} \cdot b \cdot h$. Write your answers in square units.

A.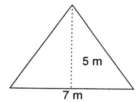

A = _____

B.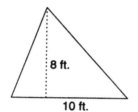

A = _____

C.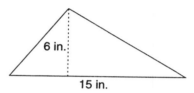

A = _____

D.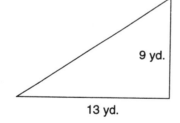

A = _____

E.

28 mi.
17 mi.

A = _____

F.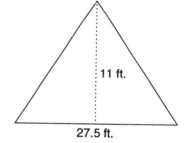

A = _____

G.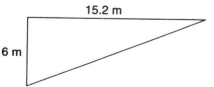

A = _____

H.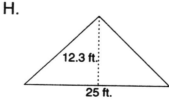

A = _____

I.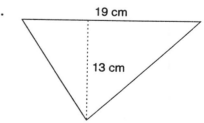

A = _____

© Frank Schaffer Publications, Inc.   FS-10218 Introduction to Geometry

Name _____ Area

Area of Trapezoids

To find the area of a trapezoid, use the formula $\frac{1}{2}$ (base₁ + base₂) • height.

$A = \frac{1}{2}(b_1 + b_2) \cdot h$

$A = \frac{1}{2}(10 + 12) \cdot 6$

$A = \frac{1}{2}(22 \cdot 6)$

$A = 66$ ft.²

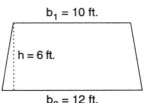

Write an equation using the formula $\frac{1}{2}(b_1 + b_2) \cdot h$. Use it to find the area of each trapezoid. Work on scratch paper. Write your answers in square units.

A.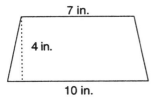

A = _____

B.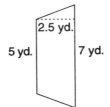

A = _____

C.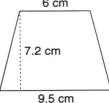

A = _____

D.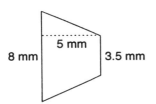

A = _____

E.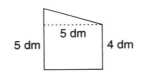

A = _____

F.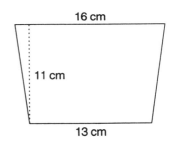

A = _____

G.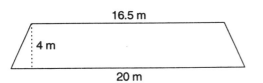

A = _____

H.

A = _____

Area of Circles

To find the area of a circle, use the formula π **r²**. Work with a calculator or on scratch paper. Round your answers to the nearest tenth. Write your answers in square units.

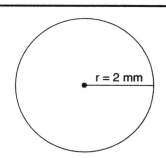

$A = \pi r^2$
$A = 3.14 \cdot (2)^2$
$A = 3.14 \cdot 4$
$A = 12.56$ mm²

A.

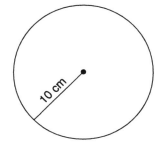 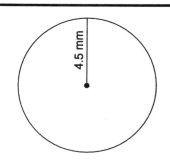 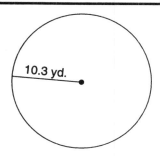

A = _____   A = _____   A = _____
A = _____   A = _____   A = _____

B.

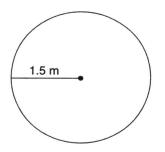 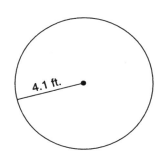 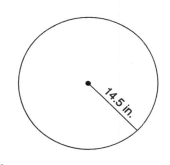

A = _____   A = _____   A = _____
A = _____   A = _____   A = _____

C.

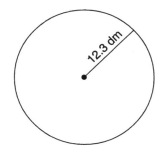 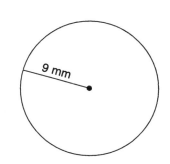 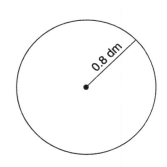

A = _____   A = _____   A = _____
A = _____   A = _____   A = _____

Name _____ Area

Mixed Practice

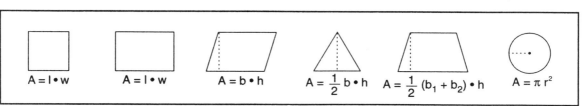

Use the correct formula to find the area of each figure below. Work with a calculator or on scratch paper. Round to the nearest tenth. Write your answers in square units.

A.
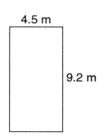

A = __4.5 • 9.2__
A = __41.4 m²__

B.
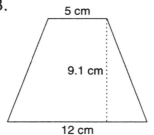

A = _____
A = _____

C.
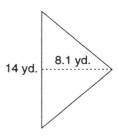

A = _____
A = _____

D.
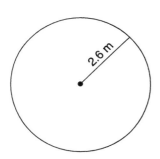

A = _____
A = _____

E.
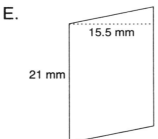

A = _____
A = _____

F.
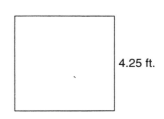

A = _____
A = _____

G.
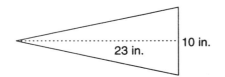

A = _____
A = _____

H.
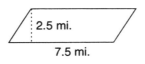

A = _____
A = _____

I.
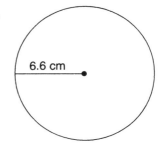

A = _____
A = _____

Name _____   Area

Total Area

Find the total area of each figure. Round your answers to the nearest tenth. Write your answers in square units.

A.

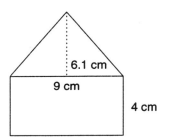

A = _____

B.

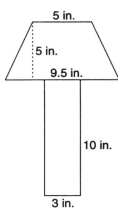

A = _____

C.

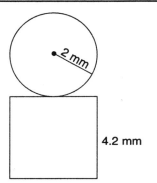

A = _____

D.

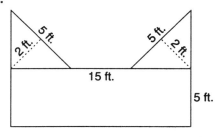

A = _____

E.

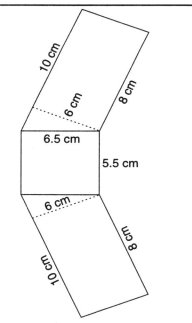

A = _____

F.

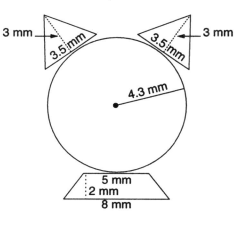

A = _____

© Frank Schaffer Publications, Inc.   46   FS-10218 Introduction to Geometry

Name _____ Coordinate graphing

Graphing in Four Quadrants

To graph an ordered pair, start at the origin, (0, 0).
Move **x** units right or left.
Then move **y** units up or down.

The ★ is at point (-1,-4). Since both numbers are negative (-), it is in Quadrant III.

To plot this point, you would
  Start at the origin.
  Move 1 unit to the left.
  Move 4 units down.

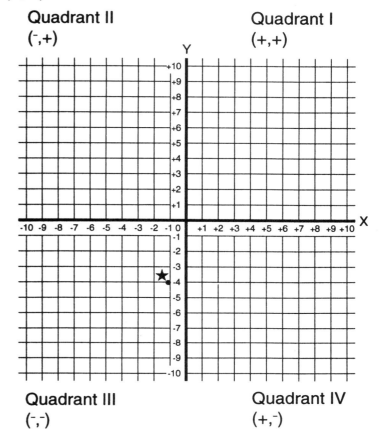

Quadrant II (-,+)
Quadrant I (+,+)
Quadrant III (-,-)
Quadrant IV (+,-)

Draw and label each point at the given location.

A. (3,4)   B. (-2,5)   C. (4,7)   D. (5,-6)

E. (-6,-8)   F. (-5,7)   G. (-4,-5)   H. (10,6)

I. (7,9)   J. (-2,-8)   K. (-10,-3)   L. (5,5)

M. (9,6)   N. (-4,-9)   O. (-9,2)   P. (8,-4)

Draw and label a point in each quadrant.
Write the location of each point.

Quadrant I   Quadrant II   Quadrant III   Quadrant IV

Q _____   R _____   S _____   T _____

© Frank Schaffer Publications, Inc.   FS-10218 Introduction to Geometry

Name _____ Segment midpoints

Finding Midpoints

$\overline{RS}$ has **endpoints** (0,1) and (8,5).

The **midpoint** of $\overline{RS}$ is the point that is halfway between the endpoints.

To calculate the midpoint of $\overline{RS}$:
• Find the the average of the x values.
$$\frac{0+8}{2} = \frac{8}{2} = 4$$
The x value of the midpoint is 4.

• Find the average of the y values.
$$\frac{1+5}{2} = \frac{6}{2} = 3$$
The y value of the midpoint is 3.

• **The midpoint of $\overline{RS}$ is (4,3).**

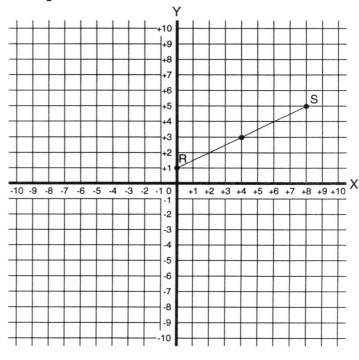

Plot the endpoints of each line segment. Draw the line segment on the graph.
Calculate the midpoint. Plot it on the graph.

Segment	Endpoint	Endpoint	Workspace	Midpoint
1. $\overline{AB}$	A (1,9)	B (3,7)		
2. $\overline{CD}$	C (-3,7)	D (-5,7)		
3. $\overline{EF}$	E (-5,-1)	F (-7,-9)		
4. $\overline{GH}$	G (4,-1)	H (8,-9)		
5. $\overline{IJ}$	I (1,6)	J (9,10)		
6. $\overline{KL}$	K (-3,-5)	L (3,-5)		
7. $\overline{MN}$	M (1,4)	N (1,-4)		
8. $\overline{OP}$	O (-10,-2)	P (-10,-10)		

© Frank Schaffer Publications, Inc.   48   FS-10218 Introduction to Geometry

Name _____ Transformations

Reflections

A **reflection** makes the mirror image of a figure.
A **line of reflection** tells how the figure is reflected.

This segment is reflected about the vertical (y) axis.

This segment is reflected about the horizontal (x) axis.

Decide whether each figure is reflected about the vertical (y) or horizontal (x) axis. Circle the x or the y.

A.

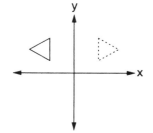

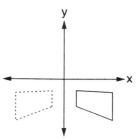

Draw a reflection of the figure about the horizontal or vertical axis as indicated.

B. vertical   horizontal   horizontal

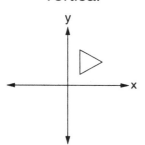

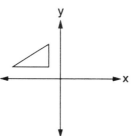

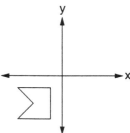

C. horizontal   vertical   vertical

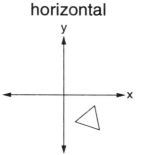

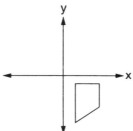

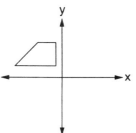

Translations and Rotations

A **translation** is the movement of an entire figure along a path without rotating or flipping.

A **rotation** is the movement of a figure about a single point.

Is the second figure a translation or a rotation of the first figure?

A.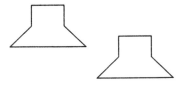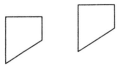

_____ _____ _____

B.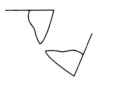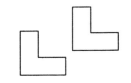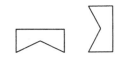

_____ _____ _____

C.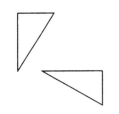

_____ _____ _____

D.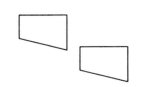

_____ _____ _____

Name _____ Symmetry

Lines of Symmetry

This triangle is folded along its **line of symmetry**.

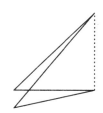

When the triangle is folded, the two parts fit exactly on top of each other.

You can see the line of symmetry when the triangle is unfolded.

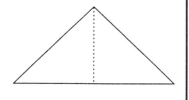

Draw what you think these figures will look like when they are unfolded. Draw a dotted line of symmetry on each unfolded figure.

A.

B.

C.

D.

E.

F.

G.

H.

© Frank Schaffer Publications, Inc.   FS-10218 Introduction to Geometry

Name _____ Symmetry

Alphabet Symmetry

> The letter B has a horizontal line of symmetry.
> The letter W has a vertical line of symmetry.

Draw and identify the lines of symmetry for the letters shown below. Write **vertical, horizontal,** or **both** to indicate the kind of symmetry. If there are no lines of symmetry, write **none.**

1. A   B   C   D   E

2. F   G   H   I   J

3. K   L   M   N   O

4. P   Q   R   S   T

5. U   V   W   X   Y

© Frank Schaffer Publications, Inc.   FS-10218 Introduction to Geometry

Name _____ Points and lines

The Point in the Middle

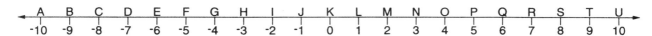

To find the midpoint of segment $\overline{GM}$, add the endpoint coordinates. Then divide by 2.

G is at -4.
M is at 2.

$\dfrac{-4 + 2}{2} = \dfrac{-2}{2} = -1$

The midpoint of $\overline{GM}$ is J at -1.

Name the location and coordinate for each segment midpoint.

1. $\overline{LP}$ _____ $\overline{KS}$ _____ $\overline{FL}$ _____

2. $\overline{AI}$ _____ $\overline{EK}$ _____ $\overline{HP}$ _____

3. $\overline{BL}$ _____ $\overline{HT}$ _____ $\overline{CM}$ _____

4. $\overline{GQ}$ _____ $\overline{DR}$ _____ $\overline{JT}$ _____

© Frank Schaffer Publications, Inc.   FS-10218 Introduction to Geometry

Name _____ Segments

Centimeter Segments

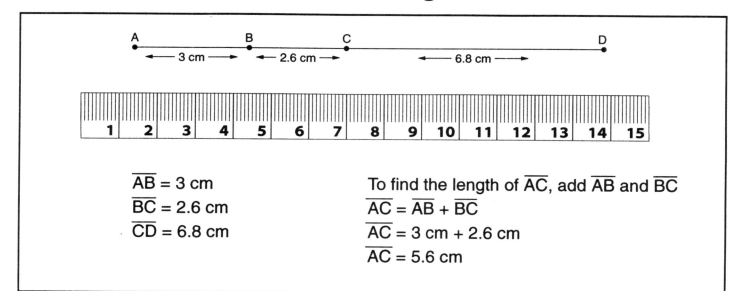

$\overline{AB}$ = 3 cm
$\overline{BC}$ = 2.6 cm
$\overline{CD}$ = 6.8 cm

To find the length of $\overline{AC}$, add $\overline{AB}$ and $\overline{BC}$
$\overline{AC} = \overline{AB} + \overline{BC}$
$\overline{AC}$ = 3 cm + 2.6 cm
$\overline{AC}$ = 5.6 cm

Use a centimeter ruler to measure the lengths.

1.

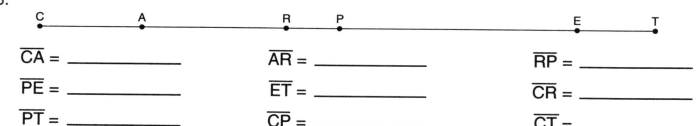

$\overline{TA}$ = _____ $\overline{AB}$ = _____ $\overline{BL}$ = _____

$\overline{LE}$ = _____ $\overline{TB}$ = _____ $\overline{AL}$ = _____

$\overline{BE}$ = _____ $\overline{TL}$ = _____ $\overline{TE}$ = _____

2.

$\overline{CH}$ = _____ $\overline{HA}$ = _____ $\overline{AI}$ = _____

$\overline{IR}$ = _____ $\overline{CA}$ = _____ $\overline{HI}$ = _____

$\overline{AR}$ = _____ $\overline{HR}$ = _____ $\overline{CR}$ = _____

3.

$\overline{CA}$ = _____ $\overline{AR}$ = _____ $\overline{RP}$ = _____

$\overline{PE}$ = _____ $\overline{ET}$ = _____ $\overline{CR}$ = _____

$\overline{PT}$ = _____ $\overline{CP}$ = _____ $\overline{CT}$ = _____

© Frank Schaffer Publications, Inc.   FS-10218 Introduction to Geometry

Constructing Line Segments

Use a centimeter ruler to construct the line segments.
Add the missing points to the segment.

Segment $\overline{CS}$ = 16 cm

C   A   S

1. $\overline{CA}$ = 4.5 cm $\overline{AR}$ = 5.6 cm $\overline{RD}$ = 3.0 cm
2. $\overline{DS}$ = _____ $\overline{CR}$ = _____ $\overline{AD}$ = _____
3. $\overline{CD}$ = _____ $\overline{AS}$ = _____ $\overline{RS}$ = _____

Segment $\overline{HT}$ = 15.2 cm

H   T

4. $\overline{HE}$ = 3.8 cm $\overline{EA}$ = 4.3 cm $\overline{AR}$ = 2.7 cm
5. $\overline{RT}$ = _____ $\overline{HA}$ = _____ $\overline{HR}$ = _____
6. $\overline{AT}$ = _____ $\overline{ER}$ = _____ $\overline{ET}$ = _____

Segment $\overline{FD}$ = 14.8 cm

F   D

7. $\overline{FI}$ = 2.7 cm $\overline{IE}$ = 3.6 cm $\overline{EL}$ = 4.2 cm
8. $\overline{LD}$ = _____ $\overline{ED}$ = _____ $\overline{ID}$ = _____
9. $\overline{FL}$ = _____ $\overline{FE}$ = _____ $\overline{IL}$ = _____

Name _____ Lines and planes

Colinear or Coplanar?

Points that lie on the same line are **colinear**.

Points that lie on the same plane (flat surface) are **coplanar**.

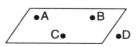

G and H are colinear.
I and G are <u>not</u> colinear.

A, B, and C are coplanar.
D and A are <u>not</u> coplanar.

Use the diagrams to complete items 1–14. A dashed line on the diagram indicates a line that goes through a plane.

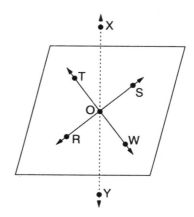

1. R, O, and _____ are colinear.
2. T, O, and _____ are colinear.
3. X, O, and _____ are colinear.
4. R, O, S, T, and _____ are coplanar.
5. R, S, and _____ are <u>not</u> coplanar.
6. X and Y are _____.

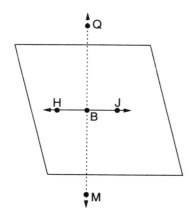

7. Draw line TV passing through point B.
8. Draw line AS that passes through point M and is not on the plane shown.
9. T, V, and _____ are colinear.
10. A, M, and _____ are colinear.
11. T, J, and _____ are coplanar.
12. H, B, and _____ are coplanar.
13. H, B, and _____ are <u>not</u> coplanar.
14. M, B, and _____ are colinear.

Name _____ Lines

Till We Meet Again

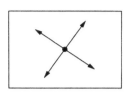
Intersecting lines meet at a point.

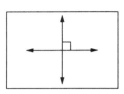
Perpendicular lines intersect and form a right angle.

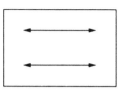
Parallel lines are in the same plane and never intersect.

Identify each pair of lines as **intersecting, perpendicular,** or **parallel.**

1. $\overleftrightarrow{GA}$ and $\overleftrightarrow{GH}$ _____
2. $\overleftrightarrow{GA}$ and $\overleftrightarrow{GE}$ _____
3. $\overleftrightarrow{HB}$ and $\overleftrightarrow{HF}$ _____
4. $\overleftrightarrow{BC}$ and $\overleftrightarrow{HF}$ _____
5. $\overleftrightarrow{HB}$ and $\overleftrightarrow{AB}$ _____
6. $\overleftrightarrow{IC}$ and $\overleftrightarrow{EC}$ _____
7. $\overleftrightarrow{EF}$ and $\overleftrightarrow{GH}$ _____
8. $\overleftrightarrow{GH}$ and $\overleftrightarrow{HB}$ _____

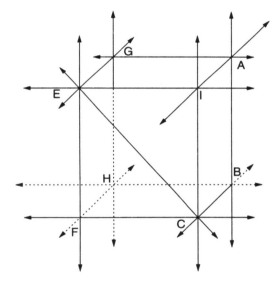

9. Name a pair of lines that are parallel. _____

10. Name a pair of lines that are perpendicular. _____

11. In the space below, draw and label a pair of intersecting lines, a pair of perpendicular lines, and a pair of parallel lines.

Name _____ Angles

All the Angles

Supplementary angles combine to make a 180° angle.	Complementary angles combine to make a 90° angle.	Adjacent angles share a vertex and a ray.	Vertical angles are not adjacent, but they share a vertex and are congruent.
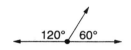	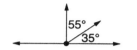	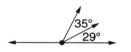	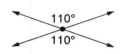

Give the complement and the supplement for each angle.

	complement	supplement			complement	supplement
A. 55°	35°	125°		27°	_____	_____
B. 68°	_____	_____		82°	_____	_____
C. 13°	_____	_____		34°	_____	_____

Use the diagram to answer the questions below.

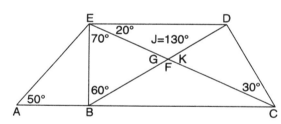

D. Which angle is adjacent to ∠DEJ? _____

E. What angle is complementary to ∠EBD? _____

F. Name two pairs of vertical angles. _____ and _____

_____ and _____

G. What is the measure of ∠DBC? _____

H. What is the measure of ∠FCB? _____

I. What is the measure of ∠BFC? _____

J. What is the measure of ∠EGB? _____

Name _____ Angles

Angle Sense

Use what you know about angles to find the missing measures.

Hint: The sum of the angles in a triangle is 180°.

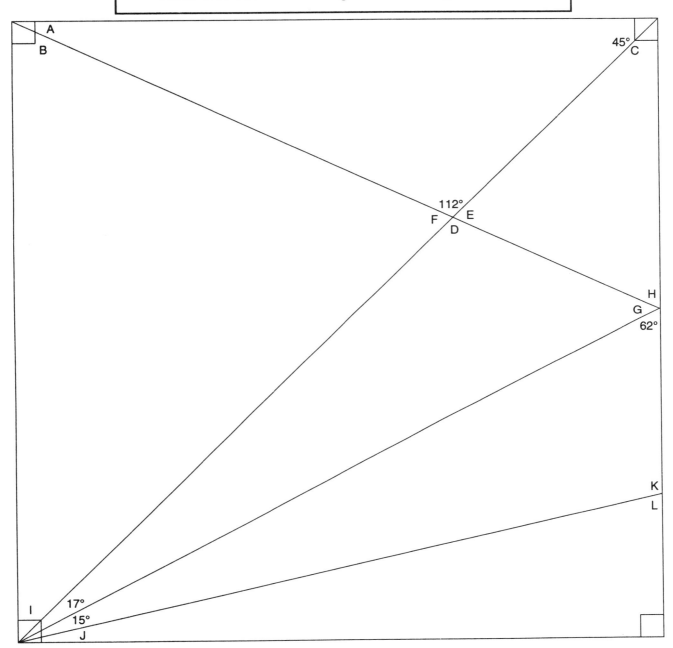

m∠A = _____   m∠B = _____   m∠C = _____   m∠D = _____

m∠E = _____   m∠F = _____   m∠G = _____   m∠H = _____

m∠I = _____   m∠J = _____   m∠K = _____   m∠L = _____

Name _____ Angles

Measuring Angles With a Protractor

Follow these steps to measure an angle using a protractor.

1. Place the center point on the vertex of the angle.
2. Position the zero edge along one side of the angle.
3. Read the hash marks to find the measure of the angle.

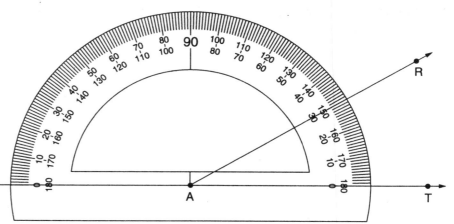

The measure of ∠RAT is 30°.

Use a protractor to find the measure of each angle.

A.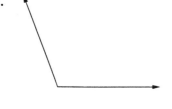

_____   _____   _____

B.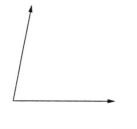

_____   _____   _____

C.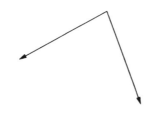

_____   _____   _____

Name _____ Angles

Dot-Paper Angles

Estimate the measure of each angle. Then use a protractor to find the actual measure. Record the estimates and the measurements.

A.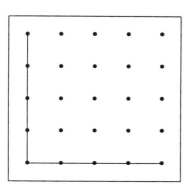

Estimate: _____

Measure: _____

B.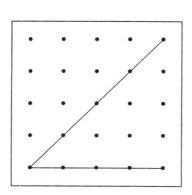

Estimate: _____

Measure: _____

C.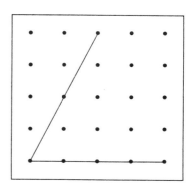

Estimate: _____

Measure: _____

D.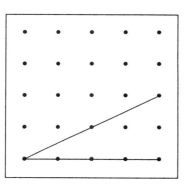

Estimate: _____

Measure: _____

E.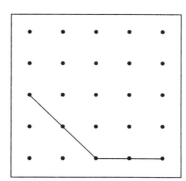

Estimate: _____

Measure: _____

F.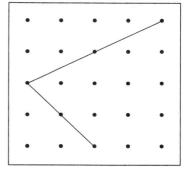

Estimate: _____

Measure: _____

G.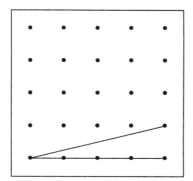

Estimate: _____

Measure: _____

H.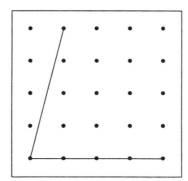

Estimate: _____

Measure: _____

I.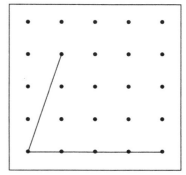

Estimate: _____

Measure: _____

© Frank Schaffer Publications, Inc.   FS-10218 Introduction to Geometry

Name _____ Angles

Time for Angles

Use a protractor to measure the angle shown on each clock face.

A.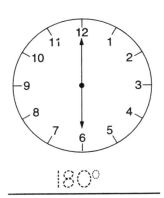

180°

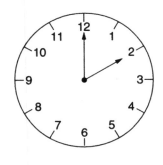

B.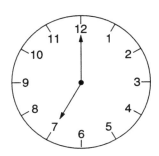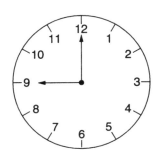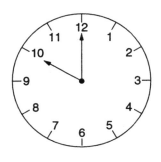

C.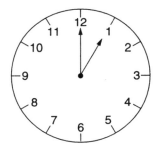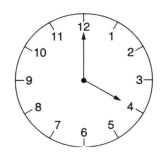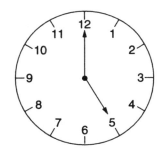

D. Name a time that is not shown that would create an angle of about 120°. _____

E. Name a time that is not shown that would create an angle of about 30°. _____

Name _____ Angles

Constructing Angles

Follow these steps to construct an angle using a protractor:
1. Draw a ray.
2. Place the center of the protractor on the endpoint and the zero mark along the ray.
3. Read the scale to the desired angle measure. Draw a point.
4. Draw a ray from the endpoint of the first ray to the new point.

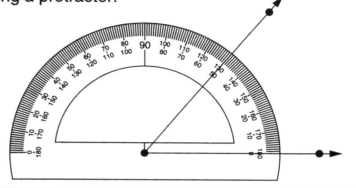

A.   90°

B.   45°

C.   60°

D.   30°

E.   120°

F.   150°

Name _____ Angles

Constructing and Classifying Angles

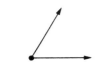    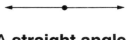

An **acute angle** is less than 90°.   A **right angle** equals 90°.   An **obtuse angle** is greater than 90° and less than 180°.   A **straight angle** equals 180°.

Use a protractor to construct angles of the given size and classify them as **acute, right, obtuse,** or **straight.**

A. 55° _____   B. 90° _____

C. 137° _____   D. 34° _____

E. 180° _____   F. 103° _____

Name _____ Angles

Alternate Angles

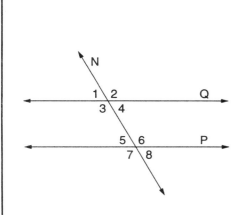

Line Q is parallel to line P.
Line N intersects lines Q and P, forming 8 angles.

Angles 4 and 5 are **alternate interior angles.**
Angles 1 and 8 are **alternate exterior angles.**

When parallel lines are intersected by the same line, the alternate interior angles are congruent and the alternate exterior angles are congruent.

Alternate interior angles
∠4 ≅ ∠5   ∠3 ≅ ∠6
Alternate exterior angles
∠1 ≅ ∠8   ∠2 ≅ ∠7

Use the diagram at the right to find the angle measures.
Line A is parallel to Line M.

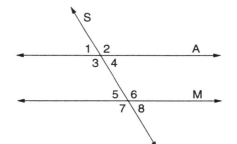

A. Let m∠1 = 58°.

m∠2 = _____   m∠3 = _____

m∠4 = _____   m∠5 = _____

m∠6 = _____   m∠7 = _____   m∠8 = _____

B. Let m∠2 = 116°.

m∠1 = _____   m∠3 = _____   m∠4 = _____   m∠5 = _____

m∠6 = _____   m∠7 = _____   m∠8 = _____

C. Let m∠5 = 63°.

m∠1 = _____   m∠2 = _____   m∠3 = _____   m∠4 = _____

m∠6 = _____   m∠7 = _____   m∠8 = _____

D. Let m∠7 = 123°.

m∠1 = _____   m∠2 = _____   m∠3 = _____   m∠4 = _____

m∠5 = _____   m∠6 = _____   m∠8 = _____

Name _____ Angles

Finding Angle Measurements

Use the information given to find the measure of each angle.

A. **X is parallel to W.**

∠1 = _____

∠2 = _____

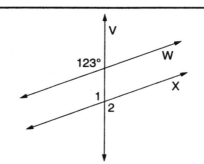

B. **Y is parallel to Z.**

∠1 = _____

∠2 = _____

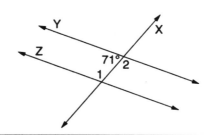

C. **B is parallel to C.**
   D is parallel to E.
   m∠1 = 53°

∠2 = _____   ∠3 = _____   ∠4 = _____

∠5 = _____   ∠6 = _____   ∠7 = _____

∠8 = _____   ∠9 = _____   ∠10 = _____

∠11 = _____   ∠12 = _____   ∠13 = _____

∠14 = _____   ∠15 = _____   ∠16 = _____

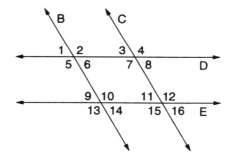

D. **R is perpendicular to U.**
   S is parallel to T.
   m∠1 = 39°

∠2 = _____   ∠3 = _____   ∠4 = _____

∠5 = _____   ∠6 = _____   ∠7 = _____

∠8 = _____   ∠9 = _____   ∠10 = _____

∠11 = _____   ∠12 = _____   ∠13 = _____

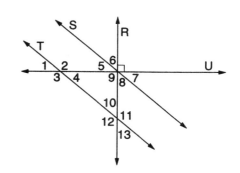

© Frank Schaffer Publications, Inc.   FS-10218 Introduction to Geometry

Ratio

A **ratio** is a comparison of two numbers. There are several ways to express a ratio.

12 out of 20   12 to 20   12:20 $\frac{12}{20} = \frac{3}{5}$

Fill in the blank spaces on the table below.

	Verbal description	a to b	a:b	$\frac{a}{b}$ (simplified)
A.	10 out of 15	10 to 15	10:15	$\frac{2}{3}$
B.	6 out of 20			
C.		9 to 10		
D.			5:50	
E.	16 out of 48			
F.				$\frac{3}{8}$
G.		13 to 52		
H.	18 out of 90			
I.			15:45	
J.	3 out of 7			
K.				$\frac{4}{11}$
L.	105 out of 150			
M.		8 to 12		

Ratio and Proportion

A proportion shows that two ratios are equal.

  $\frac{4}{12} = \frac{1}{3}$

4 out of 12 counters are shaded.
1 out of 3 rows is shaded.

Use cross products to find the missing number in a proportion.

$\frac{4}{12} \diagdown\!\!\!\!\diagup \frac{1}{n}$

$4n = 12$
$n = 3$

Solve each proportion. You may use a calculator to help you.

A. $\frac{n}{4} = \frac{6}{8}$ $\frac{n}{6} = \frac{15}{12}$ $\frac{15}{20} = \frac{n}{4}$

n = _____   n = _____   n = _____

B. $\frac{8}{36} = \frac{2}{n}$ $\frac{15}{12} = \frac{n}{4}$ $\frac{7}{8} = \frac{n}{56}$

n = _____   n = _____   n = _____

C. $\frac{7}{9} = \frac{63}{n}$ $\frac{n}{3} = \frac{15}{45}$ $\frac{14}{8} = \frac{42}{n}$

n = _____   n = _____   n = _____

D. $\frac{4}{n} = \frac{8}{3}$ $\frac{2}{n} = \frac{5}{7.5}$ $\frac{n}{39} = \frac{10}{13}$

n = _____   n = _____   n = _____

E. $\frac{7}{6} = \frac{56}{n}$ $\frac{5}{3} = \frac{105}{n}$ $\frac{n}{4.2} = \frac{1}{3}$

n = _____   n = _____   n = _____

Name _____   Similar triangles

Similar Triangles and Proportions

The symbol ~ means "is similar to." If two triangles are similar, the measurements of their corresponding sides are proportional. In the box below, △ABC ~ △DEF.

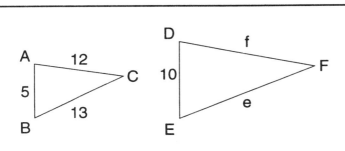

$$\frac{5}{10} = \frac{13}{e} = \frac{12}{f}$$

$$\frac{5}{10} = \frac{13}{26} = \frac{12}{24}$$

All of the corresponding sides have a ratio of 1:2.

Each pair of triangles below is similar. Set up proportions to find the missing measurements. Work on scratch paper if you need to.

1.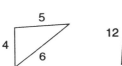

s = _____   t = _____

2.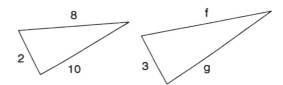

f = _____   g = _____

3.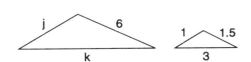

j = _____   k = _____

4.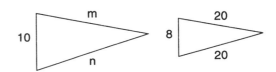

m = _____   n = _____

5.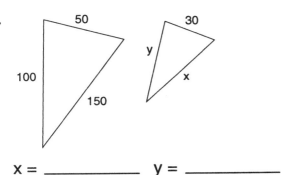

x = _____   y = _____

6.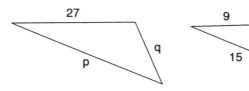

p = _____   q = _____

© Frank Schaffer Publications, Inc.   69   FS-10218 Introduction to Geometry

Name _____ Similar triangles

Proving Similarity by Angle, Angle

Triangles are similar if two angles of one triangle are congruent to two angles of another triangle. This proof is called **Angle, Angle** (AA).

90° + 42° + n = 180°   90° + 48° + m = 180°
n = 48°   m = 42°

Both triangles have angle measures of 42° and 48°.
Therefore, they are similar by Angle, Angle.

Find the missing angle measures. Write **yes** or **no** to show whether or not the triangles in each pair are similar.

1.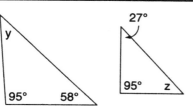

    m∠y = _____ m∠z = _____

    Similar? _____

2.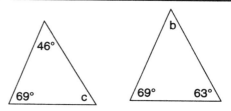

    m∠c = _____ m∠b = _____

    Similar? _____

3.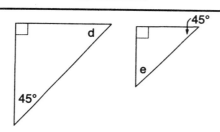

    m∠d = _____ m∠e = _____

    Similar? _____

4.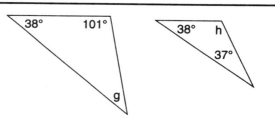

    m∠g = _____ m∠h = _____

    Similar? _____

5. If two angles of similar triangles are congruent, what do you know about the third angles? _____

© Frank Schaffer Publications, Inc.   70   FS-10218 Introduction to Geometry

Name _____ Similar triangles

Similar by Side, Side, Side or Side, Angle, Side

Triangles are similar by **Side, Side, Side** (SSS) if the sides of one triangle are proportional to the corresponding sides of the other triangle. You can cross multiply to see if the ratios are proportional.

$\frac{4}{2} = \frac{6}{3} = \frac{6}{3}$

Triangles are similar by **Side, Angle, Side** (SAS) if two sides and the angle between them of one triangle are proportional to the corresponding sides and angle of the other triangle.

$\frac{4}{2} = \frac{6}{3}$ and $90° = 90°$

Are the triangles in each pair similar? If so, write **SSS** or **SAS**. If not, write **No**.

1.

   Similar? _____

2.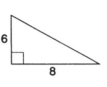

   Similar? _____

3.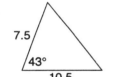

   Similar? _____

4.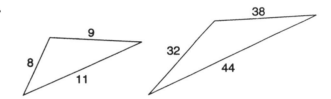

   Similar? _____

5.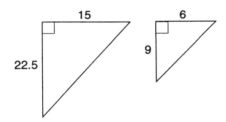

   Similar? _____

6.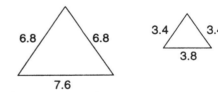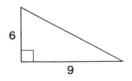

   Similar? _____

Name _____ Similar triangles

How Are They Similar?

Are the triangles in each pair similar? If so, write **AA, SSS,** or **SAS.** If not, write **No.**

1.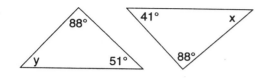

2.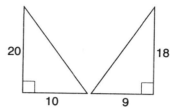

_____ _____

3.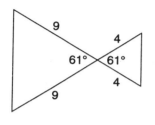

4.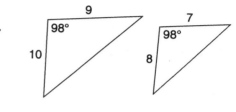

_____ _____

5.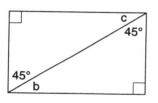

6.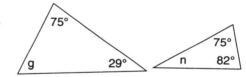

_____ _____

7.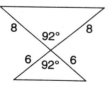

8.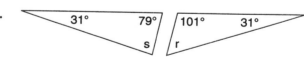

_____ _____

9.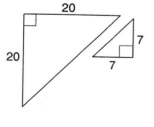

10.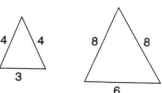

_____ _____

© Frank Schaffer Publications, Inc.   FS-10218 Introduction to Geometry

Name _____ Similar triangles

Perimeter of Similar Triangles

The perimeters of similar triangles have the same ratio as the corresponding sides of the triangles.

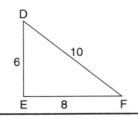

The perimeter of △ABC = 3 + 4 + 5 = 12
The perimeter of △DEF = 6 + 8 + 10 = 24
The perimeters have a ratio of $\frac{1}{2}$. The corresponding sides also have that ratio.

Find the ratio of the perimeters for each pair of similar triangles.

A.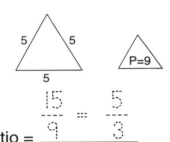

ratio = $\frac{15}{9} = \frac{5}{3}$

B.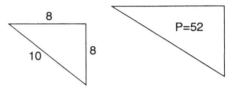

ratio = _____

C.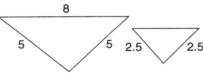

ratio = _____

D.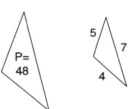

ratio = _____

E.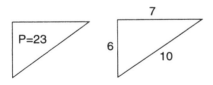

ratio = _____

F.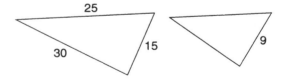

ratio = _____

G.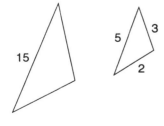

ratio = _____

H.

ratio = _____

Name _____ Squares

Squared Numbers

> To find the square of a number, multiply it by itself.
> $3^2 = 3 \cdot 3 = 9$ $(^-5)^2 = {^-5} \cdot {^-5} = 25$

Find the square of each number.

A. $3^2 =$ _____ $2^2 =$ _____ $5^2 =$ _____ $8^2 =$ _____

B. $(^-4)^2 =$ _____ $10^2 =$ _____ $6^2 =$ _____ $\left(\frac{2}{3}\right)^2 =$ _____

C. $7^2 =$ _____ $(^-1)^2 =$ _____ $\left(\frac{1}{3}\right)^2 =$ _____ $11^2 =$ _____

D. $(^-5)^2 =$ _____ $(1.1)^2 =$ _____ $13^2 =$ _____ $(^-16)^2 =$ _____

E. $\left(-\frac{2}{3}\right)^2 =$ _____ $(0.2)^2 =$ _____ $15^2 =$ _____ $\left(\frac{1}{4}\right)^2 =$ _____

F. $(0.6)^2 =$ _____ $(1.2)^2 =$ _____ $20^2 =$ _____ $(^-3)^2 =$ _____

G. $14^2 =$ _____ $(^-10)^2 =$ _____ $(0.9)^2 =$ _____ $19^2 =$ _____

H. $\left(-\frac{3}{4}\right)^2 =$ _____ $(^-20)^2 =$ _____ $(^-0.4)^2 =$ _____ $30^2 =$ _____

I. $\left(\frac{1}{5}\right)^2 =$ _____ $25^2 =$ _____ $(1.3)^2 =$ _____ $(^-0.7)^2 =$ _____

J. $(1.4)^2 =$ _____ $\left(-\frac{2}{5}\right)^2 =$ _____ $\left(-\frac{5}{6}\right)^2 =$ _____ $\left(\frac{6}{7}\right)^2 =$ _____

I. $\left(\frac{4}{5}\right)^2 =$ _____ $(^-1.4)^2 =$ _____ $(^-25)^2 =$ _____ $\left(\frac{3}{5}\right)^2 =$ _____

Square Roots

To find the **square root** of a number, find two equal factors whose product is that number. Every number has a positive square root and a negative square root.

$$-\sqrt{16}$$
$$16 = {}^-4 \cdot {}^-4$$
$$-\sqrt{16} = {}^-4$$

$$\sqrt{\frac{4}{9}}$$
$$\frac{4}{9} = \frac{2}{3} \cdot \frac{2}{3}$$
$$\sqrt{\frac{4}{9}} = \frac{2}{3}$$

Find the positive or negative square root for each number as indicated below.

A. $\sqrt{4} = \underline{2}$ $\sqrt{9} = \underline{}$ $\sqrt{25} = \underline{}$ $\sqrt{\frac{1}{4}} = \underline{}$

B. $\sqrt{\frac{1}{9}} = \underline{}$ $-\sqrt{16} = \underline{}$ $-\sqrt{49} = \underline{}$ $\sqrt{81} = \underline{}$

C. $\sqrt{0.01} = \underline{}$ $-\sqrt{36} = \underline{}$ $\sqrt{\frac{16}{25}} = \underline{}$ $\sqrt{0.36} = \underline{}$

D. $\sqrt{400} = \underline{}$ $-\sqrt{100} = \underline{}$ $\sqrt{\frac{1}{25}} = \underline{}$ $\sqrt{900} = \underline{}$

E. $\sqrt{144} = \underline{}$ $\sqrt{169} = \underline{}$ $\sqrt{0.16} = \underline{}$ $-\sqrt{\frac{9}{25}} = \underline{}$

F. $\sqrt{225} = \underline{}$ $\sqrt{\frac{81}{100}} = \underline{}$ $-\sqrt{\frac{25}{49}} = \underline{}$ $\sqrt{0.25} = \underline{}$

G. $\sqrt{196} = \underline{}$ $-\sqrt{\frac{25}{144}} = \underline{}$ $-\sqrt{1.21} = \underline{}$ $\sqrt{256} = \underline{}$

H. $-\sqrt{\frac{169}{400}} = \underline{}$ $\sqrt{\frac{49}{169}} = \underline{}$ $\sqrt{0.25} = \underline{}$ $-\sqrt{1,600} = \underline{}$

I. $\sqrt{1.44} = \underline{}$ $-\sqrt{0.81} = \underline{}$ $-\sqrt{1.69} = \underline{}$ $\sqrt{0.64} = \underline{}$

Name _____   Square roots

More About Square Roots

The numbers below are not perfect squares.
Use a calculator to find each square root.
Round your answers to the nearest hundredth.

$\sqrt{20} = 4.472136$

rounds to 4.47

A. $\sqrt{2}$ = _____ $\sqrt{10}$ = _____ $\sqrt{15}$ = _____ $\sqrt{3}$ = _____

B. $\sqrt{30}$ = _____ $\sqrt{5}$ = _____ $\sqrt{7}$ = _____ $\sqrt{99}$ = _____

C. $\sqrt{71}$ = _____ $\sqrt{17}$ = _____ $\sqrt{8}$ = _____ $\sqrt{52}$ = _____

D. $\sqrt{250}$ = _____ $\sqrt{500}$ = _____ $\sqrt{12}$ = _____ $\sqrt{33}$ = _____

E. $\sqrt{75}$ = _____ $\sqrt{150}$ = _____ $\sqrt{21}$ = _____ $\sqrt{40}$ = _____

F. $\sqrt{56}$ = _____ $\sqrt{60}$ = _____ $\sqrt{85}$ = _____ $\sqrt{90}$ = _____

G. $\sqrt{110}$ = _____ $\sqrt{125}$ = _____ $\sqrt{155}$ = _____ $\sqrt{37}$ = _____

H. $\sqrt{65}$ = _____ $\sqrt{95}$ = _____ $\sqrt{240}$ = _____ $\sqrt{525}$ = _____

I. $\sqrt{80}$ = _____ $\sqrt{600}$ = _____ $\sqrt{825}$ = _____ $\sqrt{130}$ = _____

J. $\sqrt{53}$ = _____ $\sqrt{27}$ = _____ $\sqrt{35}$ = _____ $\sqrt{1,000}$ = _____

© Frank Schaffer Publications, Inc.   FS-10218 Introduction to Geometry

Name _____ Right triangles

The Pythagorean Theorem

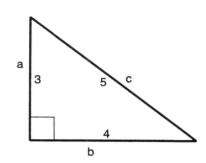

In a right triangle, the square of the length of the hypotenuse (the longest side) is equal to the sum of the squares of the lengths of the legs (the shorter sides).

In the diagram at the left, a = 3, b = 4, c = 5.

$a^2 + b^2 = c^2$
$3^2 + 4^2 = 5^2$
$9 + 16 = 25$

Use the Pythagorean Theorem to find the unknown measurement for each triangle.

A.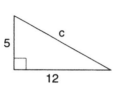

$5^2 + 12^2 = C^2$
$25 + 144 = C^2$
$C^2 = 169$
$\sqrt{C^2} = \sqrt{169}$

c = __13__

B.

c = _____

C.

c = _____

D.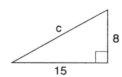

c = _____

E.

c = _____

F.

c = _____

G.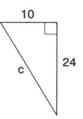

c = _____

H.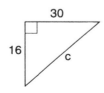

c = _____

Name _____   Right triangles

Find the Missing Measure

The Pythagorean Theorem states that the square of the hypotenuse is equal to the sum of the squares of the legs. Use the formula **$a^2+b^2=c^2$**, where a and b are the legs and c is the hypotenuse.

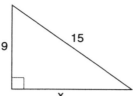

$x^2 + 9^2 = 15^2$
$x^2 + 81 = 225$
$x^2 = 225 - 81$
$x^2 = 144$
$x = \sqrt{144}$ so x = 12

Use the Pythagorean Theorem to find X. You may use a calculator if you wish.

A.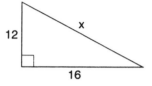

X = _____

B.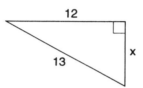

X = _____

C.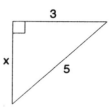

X = _____

D.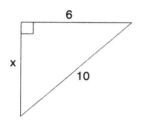

X = _____

E.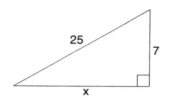

X = _____

F.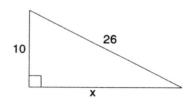

X = _____

G.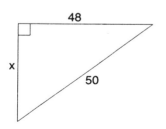

X = _____

H.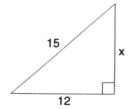

X = _____

I.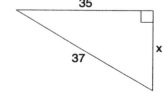

X = _____

© Frank Schaffer Publications, Inc.   FS-10218 Introduction to Geometry

Name _____   Right triangles

Use the Pythagorean Theorem

Find the value of X for each diagram. Use the formula $a^2 + b^2 = c^2$. Show each step. Use a calculator and round your answers to the nearest hundredth.

A.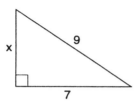

X = _____

B.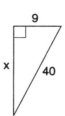

X = _____

C.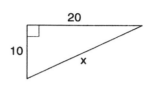

X = _____

D.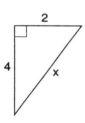

X = _____

E.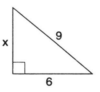

X = _____

F.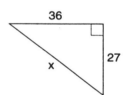

X = _____

G.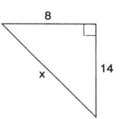

X = _____

H.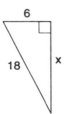

X = _____

I.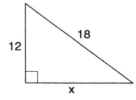

X = _____

© Frank Schaffer Publications, Inc.   FS-10218 Introduction to Geometry

Mixed Practice With Right Triangles

Make a sketch for each problem. Use the formula $a^2 + b^2 = c^2$ to solve each problem. You may use a calculator. Round your answers to the nearest hundredth and circle them. 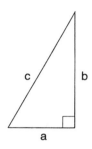	A. A right triangle has a hypotenuse of 26. The length of one leg is 15. What is the length of the other leg?
B. A triangle has a short leg with a length of 9. The other leg is twice as long. How long is the hypotenuse?	C. The legs of a right triangle measure 12 and 16. What is the length of the hypotenuse?
D. The hypotenuse of a right triangle is 17. One leg measures 15. What is the length of the other leg?	E. A triangular sail is 82 feet high. Its width is 29 feet. What is the length of the sail's hypotenuse?
F. A 25-foot ladder is leaning against a wall. It forms the hypotenuse of a right triangle. The bottom of the ladder is 6 feet from the wall. How far up the wall will the ladder reach?	G. Both legs of a right triangle measure 3. What is the length of the hypotenuse?

Name _____ Polyhedrons

Polyhedron Nets

A polyhedron is a solid figure in which all the sides are polygons. A polyhedron's **net** is made by unfolding the polyhedron so that all of its faces are visible.

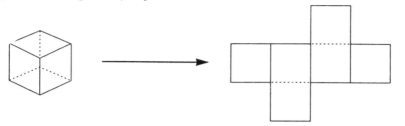

6 square faces

A solid cube makes this **net** when it is unfolded. The net has 6 sides. Each side is a square.

Match each solid with its net and a description of its faces.

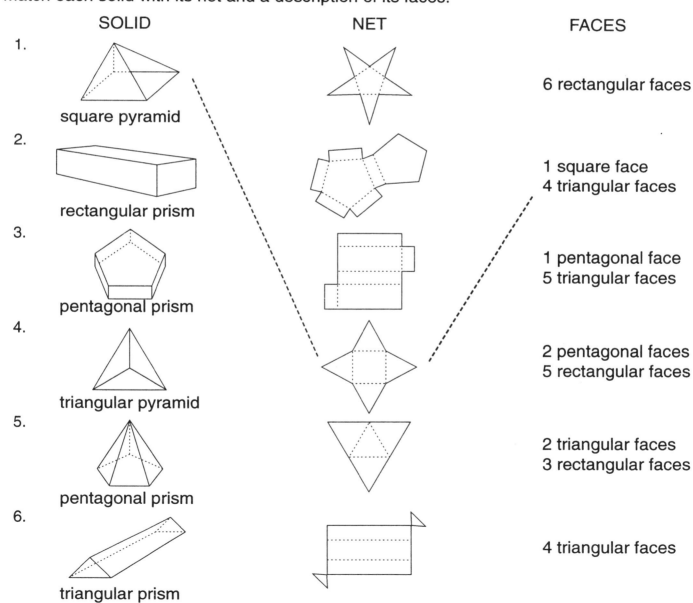

Name _____ Polyhedrons

Pyramids and Prisms

A polyhedron is a solid figure in which all the sides are polygons. Pyramids and prisms are polyhedrons.

A **pyramid** has one base. All the other faces meet at a single point.

 This is a square pyramid because the base is a square.

A **prism** has two congruent bases.

 This is a rectangular prism because the bases are rectangles.

Identify each polyhedron as a **pyramid** or a **prism**. Use the shape of the base as your label. (The bases shown are triangular, rectangular, pentagonal, hexagonal, or octagonal.)

A.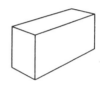

_____ _____ _____

B.

_____ _____ _____

C.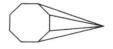

_____ _____ _____

D. Why aren't cylinders, cones, and spheres polyhedrons?

_____

Polyhedron Parts

All polyhedrons have bases, faces, edges, and vertices. Two faces on a polyhedron meet at an **edge.** Three or more edges meet at a **vertex.**

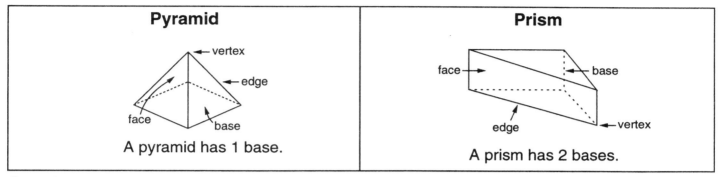

A pyramid has 1 base. A prism has 2 bases.

Complete the table. Write the number of bases. Find the missing numbers by using the formula **F + V = E + 2,** where F is the number of faces, V is the number of vertices, and E is the number of edges.

	Polyhedron name	Number of bases	Number of faces	Number of vertices	Number of edges
A.	triangular prism		5	6	
B.	triangular pyramid			4	6
C.	rectangular prism		6		12
D.	rectangular pyramid			5	8
E.	pentagonal prism		7	10	
F.	pentagonal pyramid		6		10
G.	hexagonal prism			12	18
H.	hexagonal pyramid		7	7	
I.	octagonal prism		10		24
J.	octagonal pyramid		9	9	

Name _____   Surface area

Surface Area of a Cube

A cube is made up of 6 congruent faces. Each face is a square.

 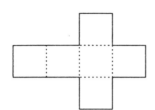

To find the **surface area (SA)** of a cube, find the area of one face. Then multiply by 6.

For the diagram above, find the area of one face by multiplying 4 • 4.
SA = 6(4 • 4)
SA = 96 u²

Find the surface area of each cube. Write an equation to help you. Round your answers to the nearest whole number.

A.

SA = __6(7•7)__
SA = __294 u²__

B.

SA = _____
SA = _____

C.

SA = _____
SA = _____

D.

SA = _____
SA = _____

E.

SA = _____
SA = _____

F.

SA = _____
SA = _____

Name _____ Surface area

Surface Area of a Rectangular Prism

To find the surface area (SA) of a rectangular prism, find the area of each face. Then add the areas to find the total.

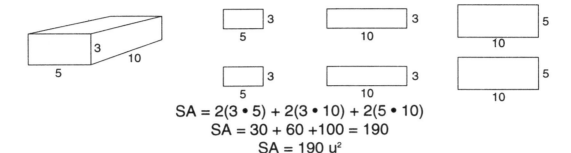

SA = 2(3 • 5) + 2(3 • 10) + 2(5 • 10)
SA = 30 + 60 + 100 = 190
SA = 190 u²

Find the surface area of each rectangular prism. Write your answer in square units (u²).

A.

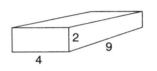

SA = _____

B.

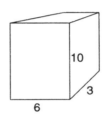

SA = _____

C.

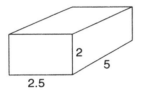

SA = _____

D.

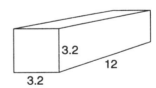

SA = _____

E.

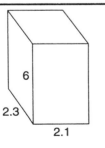

SA = _____

F.

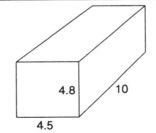

SA = _____

© Frank Schaffer Publications, Inc.   85   FS-10218 Introduction to Geometry

Name _____   Surface area

Surface Area of a Square Pyramid

To find the surface area of a square pyramid, find the total area of the four congruent faces and the area of the base. Then add the area of the faces to the area of the base.

 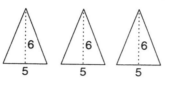 

Step #1 SA of the Faces	Step #2 SA of the Base	Step #3 SA of Square Pyramid
$SA = 4 \cdot (\frac{1}{2} \cdot b \cdot h)$	$SA = b^2$	$SA = 4 \cdot (\frac{1}{2} \cdot b \cdot h) + b^2$
$SA = 4 \cdot \frac{1}{2} \cdot 5 \cdot 6$	$SA = 5 \cdot 5$	$SA = 4 \cdot \frac{1}{2} \cdot 5 \cdot 6 + (5 \cdot 5)$
$SA = 60\ u^2$	$SA = 25\ u^2$	$SA = 60\ u^2 + 25\ u^2$
		$SA = 85\ u^2$

Find the surface area of each pyramid. Write your answers in square units (u^2.)

A.
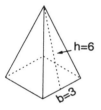

SA = _____

B.
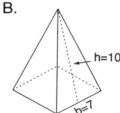

SA = _____

C.
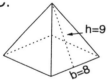

SA = _____

D.
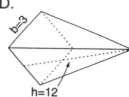

SA = _____

E.
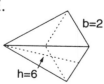

SA = _____

F.
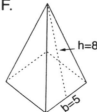

SA = _____

Name _____ Surface area

Surface Area of a Cylinder

To find the surface area (SA) of a cylinder, find the area of the curved surface and add it to the total area of both bases. To do this, use the formula **SA = 2πrh + 2πr²**.

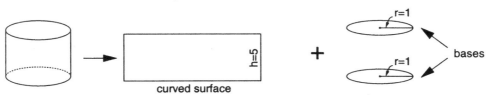

Step #1 SA of the Curved Surface	Step #2 SA of the Bases	**Step #3 SA of the Cylinder**
SA = 2πrh	SA = 2πr²	SA = 2πrh + 2πr²
SA = 2 • π • radius • height	SA = 2 • π • radius²	SA = 31.4 u² + 6.28 u²
SA = 2 • 3.14 • 1 • 5	SA = 2 • 3.14 • 1 • 1	SA = 37.68 u²
SA = 31.4 u²	SA = 6.28 u²	

Use the formula **SA = 2πrh + 2πr²** to find the surface area of each cylinder.

A.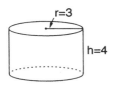

SA = _____

B.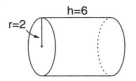

SA = _____

C.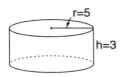

SA = _____

D.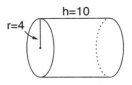

SA = _____

E.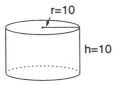

SA = _____

Name _____   Surface area

Surface Area: Mixed Practice

Find the surface area of each solid. You may use a calculator to help you.

A.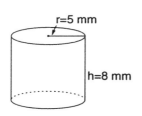

SA = _____

B.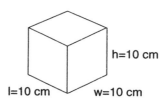

SA = _____

C.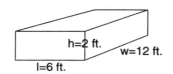

SA = _____

D.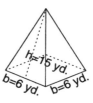

SA = _____

E.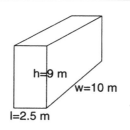

SA = _____

F.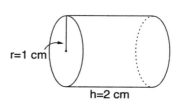

SA = _____

G.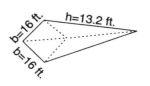

SA = _____

H.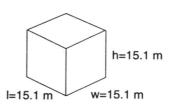

SA = _____

Name _____ Volume

Volume of a Rectangular Prism

To find the volume (V) of a rectangular prism, use the formula **Volume = Base • height**, where B is the area of the base and h is the height.

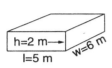

Step #1 Find the Area of the Base	Step #2 Find the Volume
B = length • width	V = Base • height
B = 5 m • 6 m	V = 30 m² • 2 m
B = 30 m²	V = 60 m³

Find the volume of each rectangular prism using the formula **V = Bh**. Express volume in cubic units.

A.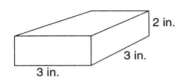

_____

B.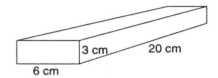

_____

C.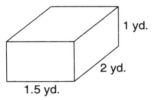

_____

D.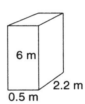

_____

E.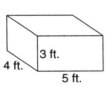

_____

F.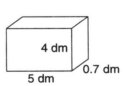

_____

G.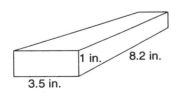

_____

H.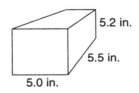

_____

© Frank Schaffer Publications, Inc. 89 FS-10218 Introduction to Geometry

Name _____ Volume

Volume of a Cube

To find the volume of a cube, use the formula **Volume = Base • height,** where B is the area of the base and h is the height.

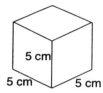

Step #1 Find the Area of the Base	Step #2 Find the Volume
B = length • width	V = Base • height
B = 5 • 5	V = 25 cm² • 5 cm
B = 25 cm²	V = 125 cm³

Find the volume of each cube. Since the length, width, and height of a cube are equal, you can use the formula **V = S³**. Round your answers to the nearest hundredth.

A.
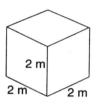
V = _____

B.

V = _____

C.
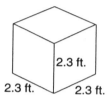
V = _____

D.
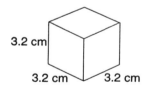
V = _____

E.

V = _____

F.
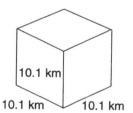
V = _____

G.
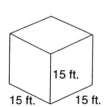
V = _____

H.
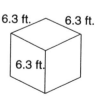
V = _____

I.
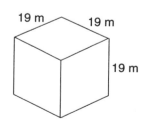
V = _____

Volume of a Triangular Prism

To find the volume of a triangular prism, use the formula **Volume = B • h**, where B is the area of the base of the prism and h is the height of the prism. The base of a triangular prism is a triangle.

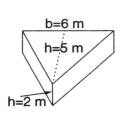

Step #1 Area of the Base	Step #2 Volume
Base = $\frac{1}{2}$ • base of triangle • height of triangle	Volume = Base • height
Base = $\frac{1}{2}$ • 6 m • 5 m	Volume = 15 m² • 2 m
Base = $\frac{1}{2}$ • 30 m²	Volume = 30 m³
Base = 15 m²	

Find the volume.

A.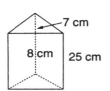

V = _____

B.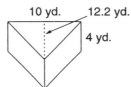

V = _____

C.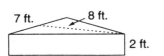

V = _____

D.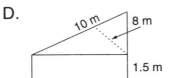

V = _____

E.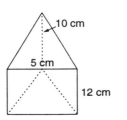

V = _____

F.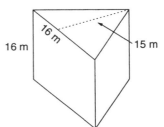

V = _____

© Frank Schaffer Publications, Inc.   FS-10218 Introduction to Geometry

Name _____ Volume

Volume of a Pyramid

To find the volume of a pyramid, use the formula **Volume = $\frac{1}{3}$ • Base • height,** where B is the area of the base and h is the height of the pyramid.

Step #1 Find the Area of the Base	Step #2 Find the Volume of the Pyramid
If the base is a rectangle, use the formula **B = length • width.**	V = $\frac{1}{3}$ • Base • height of the Pyramid
If the base is a triangle, use the formula **B = $\frac{1}{2}$ • b • h.**	

Find the volume of each pyramid. Use the formula V = $\frac{1}{3}$ • Base • height

A.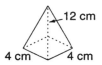

V = _____

B.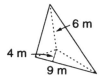

V = _____

C.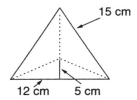

V = _____

D.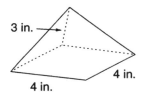

V = _____

E.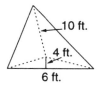

V = _____

F.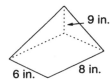

V = _____

G.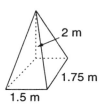

V = _____

H.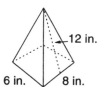

V = _____

© Frank Schaffer Publications, Inc. FS-10218 Introduction to Geometry

Name _____ Volume

Volume of a Cylinder

To find the volume of a cylinder, use the formula **Volume = Base • height,** where B is the area of the base and h is the height.

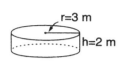

Step #1 Find the Area of the Base	Step #2 Find the Volume
$B = \pi r^2$	Volume = Base • height
$B = 3.14 \cdot 3\ m^2$	Volume = $28.26\ m^2 \cdot 2\ m$
$B = 28.26\ m^2$	Volume = $56.52\ m^3$

Find the volume of each cylinder. Use the formula **V = B • h.** Use a calculator to help you. Round your answers to the nearest hundredth.

A.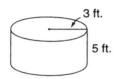

V = _____

B.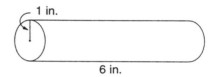

V = _____

C.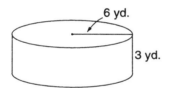

V = _____

D.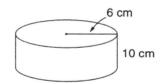

V = _____

E.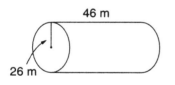

V = _____

F.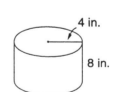

V = _____

G.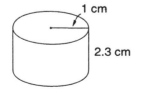

V = _____

H.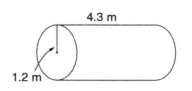

V = _____

© Frank Schaffer Publications, Inc.   FS-10218 Introduction to Geometry

Name _____ Volume

Volume of a Cone

To find the volume of a cone, use the formula **Volume = $\frac{1}{3}$ • Base • height,** where B is the area of the base and h is the height of the cone.

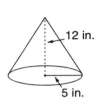

Step #1 Find the Area of the Base	**Step #2 Find the Volume**
$B = \pi \cdot r^2$	$V = \frac{1}{3}$ • Base • height
$B = 3.14 \cdot 5 \text{ in.} \cdot 5 \text{ in.}$	$V = \frac{1}{3} \cdot 78.5 \text{ in.}^2 \cdot 12 \text{ in.}$
$B = 3.14 \cdot 25 \text{ in.}^2$	
$B = 78.5 \text{ in.}^2$	$V = 314 \text{ in.}^3$

Find the volume of each cone. Use the formula **V = B • h.** Use a calculator to help you. Round your answers to the nearest hundredth.

A.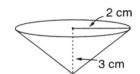

V = _____

B.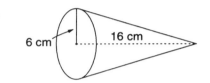

V = _____

C.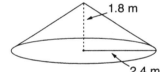

V = _____

D.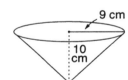

V = _____

E.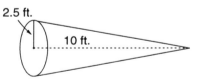

V = _____

F.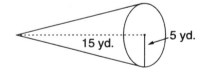

V = _____

G.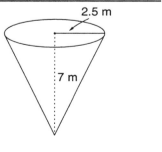

V = _____

H.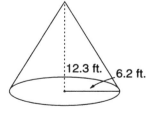

V = _____

© Frank Schaffer Publications, Inc.   FS-10218 Introduction to Geometry

Name _____ Volume

Volume: Mixed Practice

Find the volume of each figure. You may use a calculator. Round your answers to the nearest hundredth.
The volume of a prism or cylinder is **Base • height**.
The volume of a pyramid or cone is $\frac{1}{3}$ • **Base • height**.
Remember! **Base** means the area of the base.

A.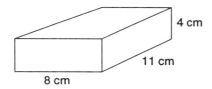

V = _____

B.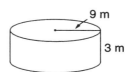

V = _____

C.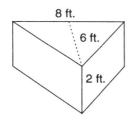

V = _____

D.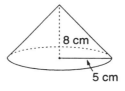

V = _____

E.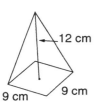

V = _____

F.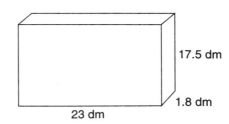

V = _____

G.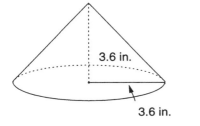

V = _____

H.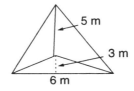

V = _____

© Frank Schaffer Publications, Inc.   FS-10218 Introduction to Geometry

Name _____ Volume

Total Volume

To find the total volume for the figures pictured, find the volume of each solid figure that makes up the figure. Then add the volumes. You may use a calculator. Record each formula that you use and the total volume in the space beside the diagram.

A.

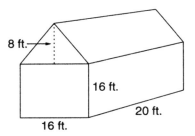

V = _____

B.

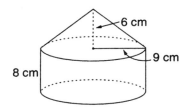

V = _____

C.

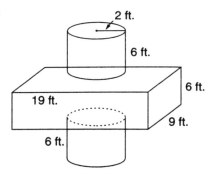

V = _____

D.

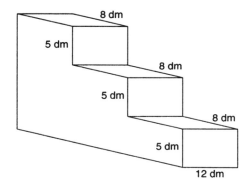

V = _____

© Frank Schaffer Publications, Inc.

Graphing Points and Lines

A line can be graphed from as few as two points. The line shown connects the points (1,1) and (2,3).

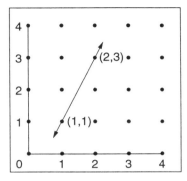

Plot the points on each dot diagram. Then draw a line that connects the points.

A. (0,0)   (3,4)   (1,2)   (3,4)   (2,1)   (4,4)

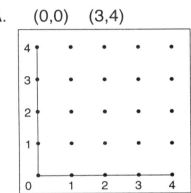   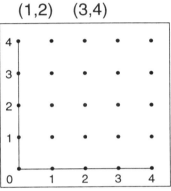   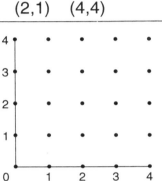

B. (3,3)   (3,4)   (1,2)   (1,4)   (1,3)   (4,3)

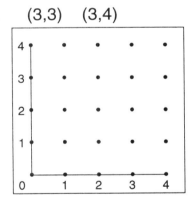   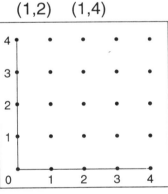   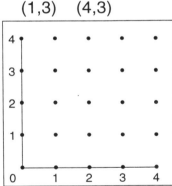

C. (1,4)   (4,1)   (4,3)   (0,2)   (1,0)   (0,4)

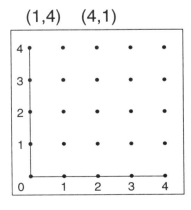   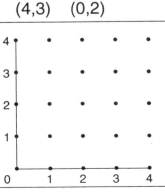   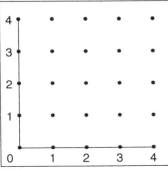

Graphing Lines on a Coordinate Plane

A line can be graphed from any two points on a coordinate plane. Points (0,⁻1) and (5,6) are plotted on the graph at the right. The points are connected by a line.

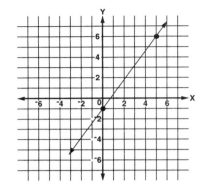

Plot the points on each coordinate plane. Draw a line that connects the points.

A. (3,5) (4,6)   (⁻2,⁻2) (2,2)   (1,5) (⁻1,⁻3)

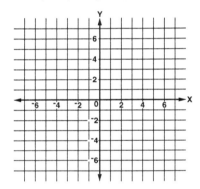 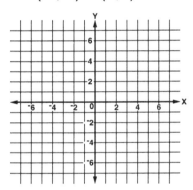 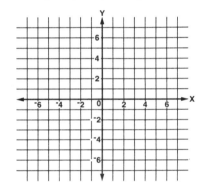

B. (5,1) (1,5)   (⁻3,4) (0,0)   (3,⁻1) (0,⁻5)

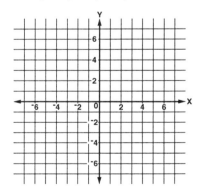 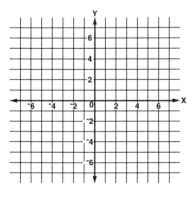 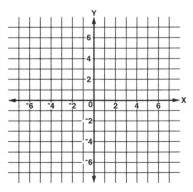

Name _____ Slope

Using a Ratio to Find Slope

You can describe the **slope** (steepness) of a line by using the ratio of the **rise** to the **run** of any two points on the line.

To find the rise: Count the units up or down between the points. (In the diagram at the right, the rise is 2 units.)

To find the run: Count the units right or left between the points. (In the diagram at the right, the run is 6 units.)

slope = $\frac{rise}{run} = \frac{2}{6} = \frac{1}{3}$

The slope of line X = $\frac{1}{3}$

Write a ratio to describe the slope of each line.

A.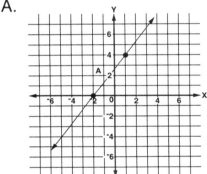

slope = _____

B.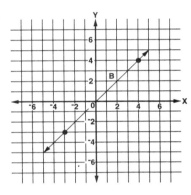

slope = _____

C.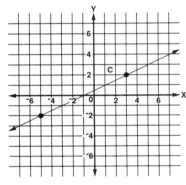

slope = _____

D.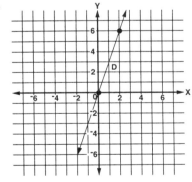

slope = _____

E.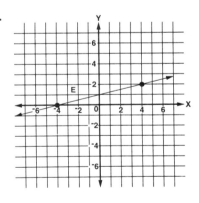

slope = _____

F.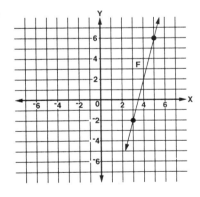

slope = _____

G. How are slopes that are greater than one different from slopes that are less than one?

_____

© Frank Schaffer Publications, Inc.   FS-10218 Introduction to Geometry

Name _____ Slope

Special Slopes

Some lines have special characteristics and slopes.

A horizontal line has a slope of 0.	**A vertical line has an undefined slope.**	**A diagonal line has a slope of 1.**
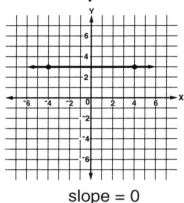 slope = 0	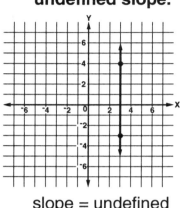 slope = undefined	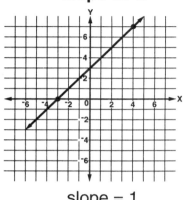 slope = 1

Plot the points on each coordinate plane. Draw a line that connects the points. Then identify the slope as **0, undefined,** or **1**.

A. (2,5) (2,1)

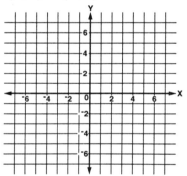

slope = _____

B. (1,4) (6,4)

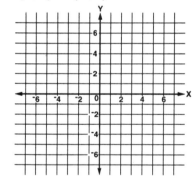

slope = _____

C. (3,3) (6,6)

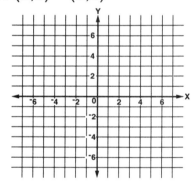

slope = _____

D. (-1,4) (-3,2)

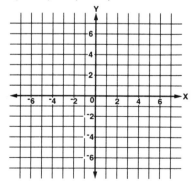

slope = _____

E. (-3,-3) (-3,2)

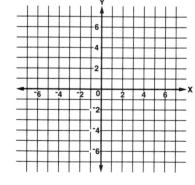

slope = _____

F. (1,-4) (-5,-4)

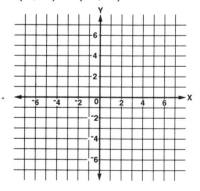

slope = _____

© Frank Schaffer Publications, Inc.   FS-10218 Introduction to Geometry

Name _____ Slope

Estimating Slope

You can tell whether a line has a positive or negative slope just by looking at it. Look at the point of the line that is the farthest to the left on the graph. If the line moves upward from that point, it has a positive slope. If the line moves downward from that point, it has a negative slope.

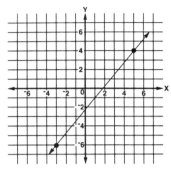

This line slants up from the lower left. That means the slope is positive.

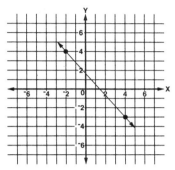

This line slants down from the upper left. That means the slope is negative.

Study each line. Circle **positive** or **negative** to show its slope. Then use the ratio formula **slope** = $\frac{\text{rise}}{\text{run}}$ to find the actual slope.

A.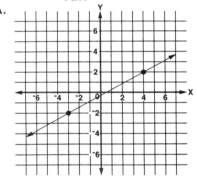

Positive   Negative

Actual: _____

B.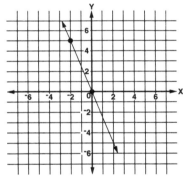

Positive   Negative

Actual: _____

C.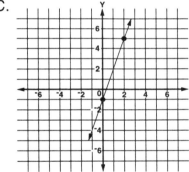

Positive   Negative

Actual: _____

D.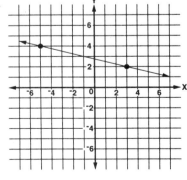

Positive   Negative

Actual: _____

E.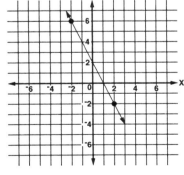

Positive   Negative

Actual: _____

F.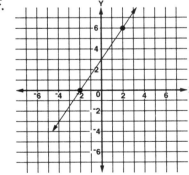

Positive   Negative

Actual: _____

Name _____   Slope

Using Coordinate Points to Find Slope

Use any two points on a line to find its slope. Use the ratio formula $\dfrac{y_2 - y_1}{x_2 - x_1}$.

x_1 and x_2 are the first coordinates of two points on a line.
y_1 and y_2 are the second coordinates of the same two points.
The points on the line shown at the right are (5,4) and (6,-1).

$x_1 = 5 \qquad x_2 = 6 \qquad y_1 = 4 \qquad y_2 = {}^-1$

The slope $= \dfrac{y_2 - y_1}{x_2 - x_1} = \dfrac{{}^-1 - 4}{6 - 5} = \dfrac{{}^-5}{1} = {}^-5$

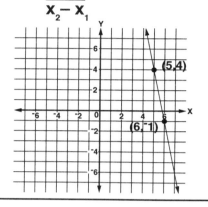

Use the formula $\dfrac{y_2 - y_1}{x_2 - x_1}$ to find the slope of each line.

A. (6,3)   (2,0)

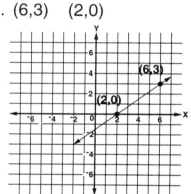

slope = _____

B. (-4,2)   (-6,5)

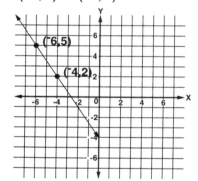

slope = _____

C. (-6,3)   (-4,5)

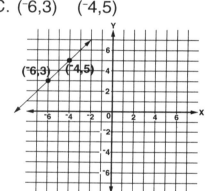

slope = _____

D. (-2,-3)   (0,1)

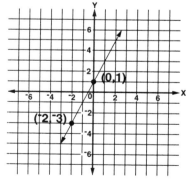

slope = _____

E. (0,3)   (6,0)

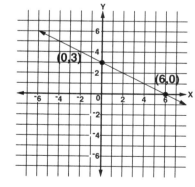

slope = _____

F. (0,-3)   (3,-1)

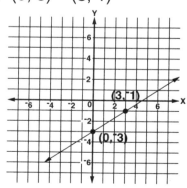

slope = _____

© Frank Schaffer Publications, Inc.   102   FS-10218 Introduction to Geometry

Name _____ Slope

Finding Slope With Coordinate Points

Plot the points on each graph. Draw a line to connect them. Then use the ratio formula **slope** = $\frac{\text{rise}}{\text{run}}$ or the point formula $\frac{y_2-y_1}{x_2-x_1}$ to find the actual slope. Identify the slope of each line as **0, 1, -1,** or **other**.

A. (3,4) (-1,0)

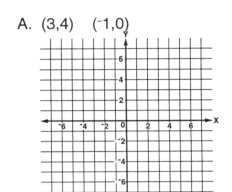

Slope: _____

B. (5,7) (2,1)

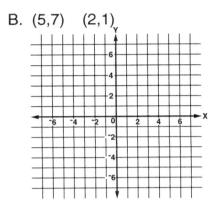

Slope: _____

C. (-3,0) (0,-4)

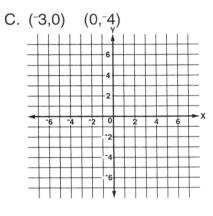

Slope: _____

D. (5,-1) (-3,0)

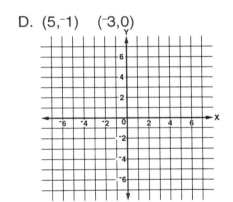

Slope: _____

E. (1,4) (-3,4)

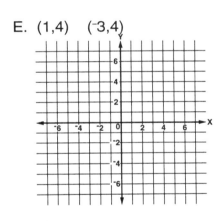

Slope: _____

F. (0,0) (2,5)

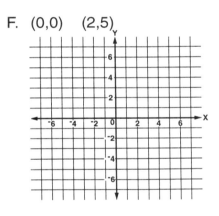

Slope: _____

G. (5,-5) (-1,-5)

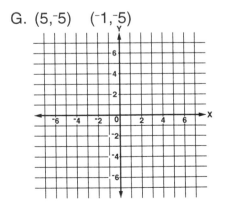

Slope: _____

H. (-4,-4) (0,0)

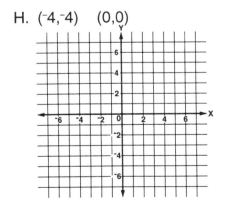

Slope: _____

I. (6,3) (2,-1)

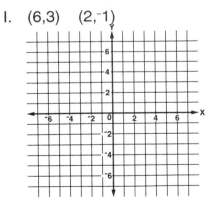

Slope: _____

© Frank Schaffer Publications, Inc. 103 FS-10218 Introduction to Geometry

Name _____ Slope

Representing Lines With Equations

You can describe a line using the equation **y = mx + b** where m is the slope and b is the y-intercept. (The y-intercept is the point where the line intersects the y-axis.)

> The line graphed at the right has a slope of 2.
> The y-intercept is 4.
> The equation for the line is **y = 2x + 4.**

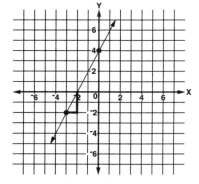

Look at each line. Use the slope and y-intercept to complete the equation for each line.

A. slope = 3
   y-int = -5
   y = _____

B. slope = -2
   y-int = 4
   y = _____

C. slope = 1
   y-int = 3
   y = _____

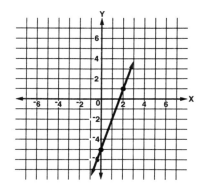
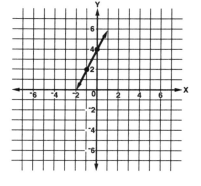
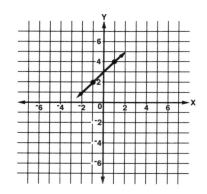

D. slope = 2
   y-int = -3
   y = _____

E. slope = -1
   y-int = 5
   y = _____

F. slope = 3
   y-int = 3
   y = _____

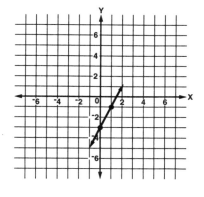
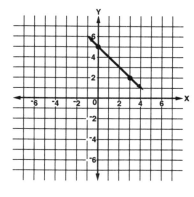
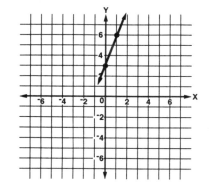

Answer Key

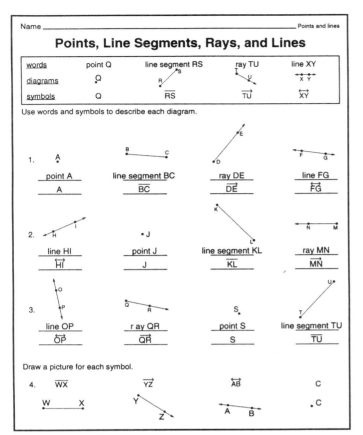

Page 1

Segment Lengths and Midpoints

	Segment	Endpoints	Length	Midpoint
1.	$\overline{KM}$	3, 5	2	4
2.	$\overline{OW}$	7, 15	8	11
3.	$\overline{RZ}$	10, 18	8	14
4.	$\overline{IS}$	1, 11	10	6
5.	$\overline{PT}$	8, 12	4	10
6.	$\overline{LZ}$	4, 18	14	11
7.	$\overline{MO}$	5, 7	2	6
8.	$\overline{JP}$	2, 8	6	5
9.	$\overline{TX}$	12, 16	4	14
10.	$\overline{NV}$	6, 14	8	10
11.	$\overline{HX}$	0, 16	16	8
12.	$\overline{HZ}$	0, 18	18	9

Page 2

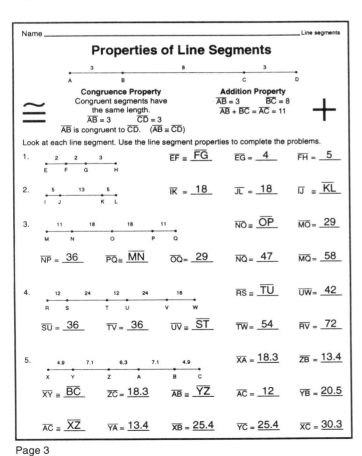

Page 3

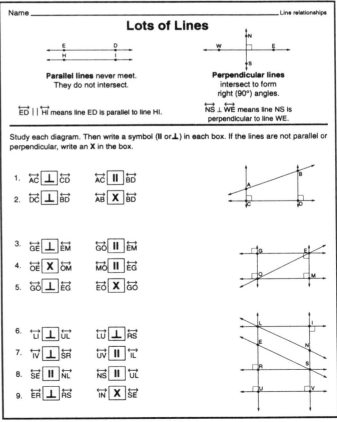

Page 4

105

© Frank Schaffer Publications, Inc. FS-10218 Introduction to Geometry

Answer Key

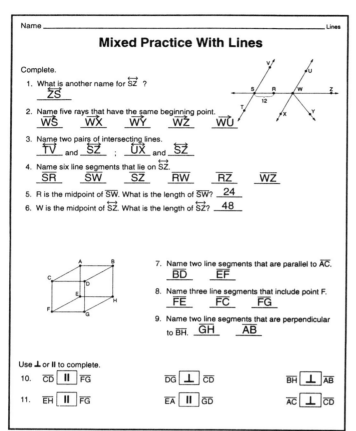

Page 5

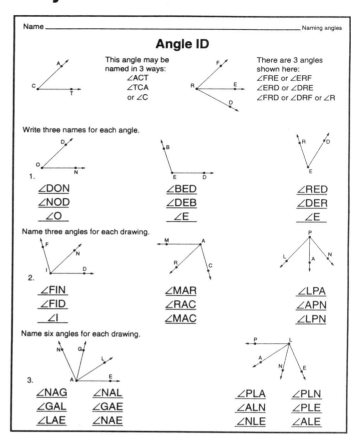

Page 6

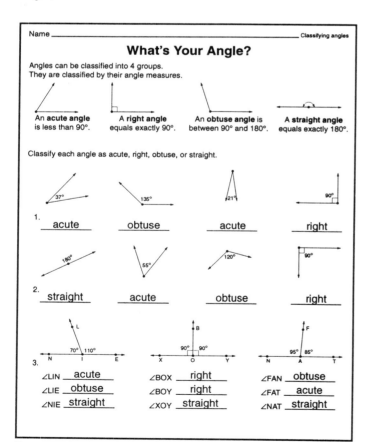

Page 7

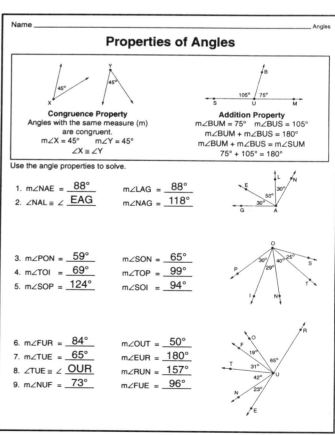

Page 8

Answer Key

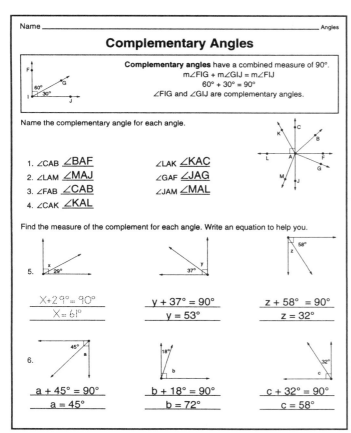

Page 9

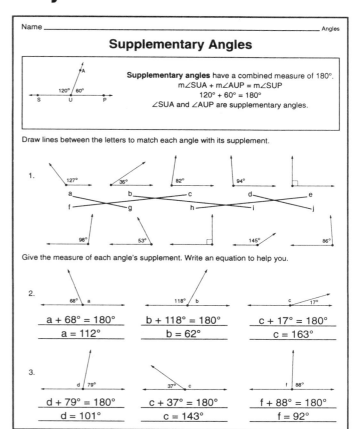

Page 10

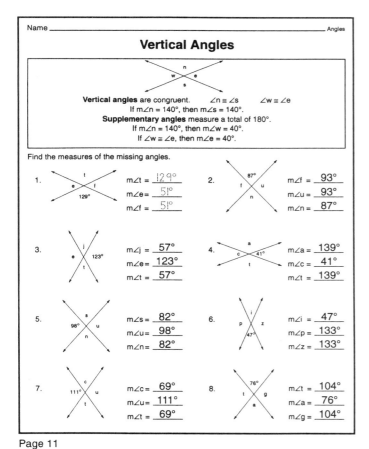

Page 11

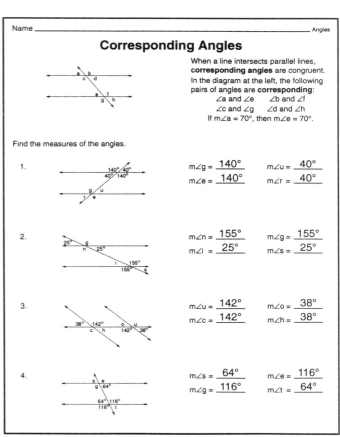

Page 12

Answer Key

Finding Angle Measures

Find the measures of the angles.
Use what you know about complementary, supplementary, and vertical angles to help you.

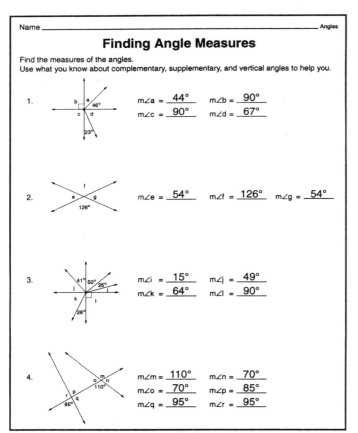

1. m∠a = __44°__   m∠b = __90°__
   m∠c = __90°__   m∠d = __67°__

2. m∠e = __54°__   m∠f = __126°__   m∠g = __54°__

3. m∠i = __15°__   m∠j = __49°__
   m∠k = __64°__   m∠l = __90°__

4. m∠m = __110°__   m∠n = __70°__
   m∠o = __70°__   m∠p = __85°__
   m∠q = __95°__   m∠r = __95°__

Page 13

More Angle Measures

Find the measure of each angle.
Use what you know about supplementary, vertical, and corresponding angles to help you.

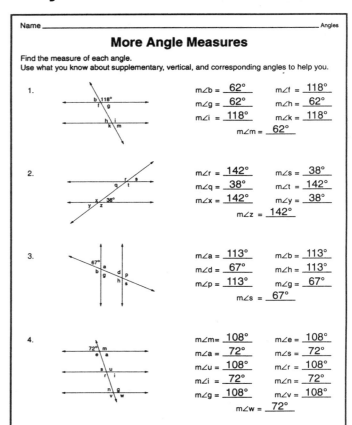

1. m∠b = __62°__   m∠f = __118°__
   m∠g = __62°__   m∠h = __62°__
   m∠i = __118°__   m∠k = __118°__
   m∠m = __62°__

2. m∠r = __142°__   m∠s = __38°__
   m∠q = __38°__   m∠t = __142°__
   m∠x = __142°__   m∠y = __38°__
   m∠z = __142°__

3. m∠a = __113°__   m∠b = __113°__
   m∠d = __67°__   m∠h = __113°__
   m∠p = __113°__   m∠g = __67°__
   m∠s = __67°__

4. m∠m = __108°__   m∠e = __108°__
   m∠a = __72°__   m∠s = __72°__
   m∠u = __108°__   m∠r = __108°__
   m∠i = __72°__   m∠n = __72°__
   m∠g = __108°__   m∠v = __108°__
   m∠w = __72°__

Page 14

Classifying Triangles by Their Sides

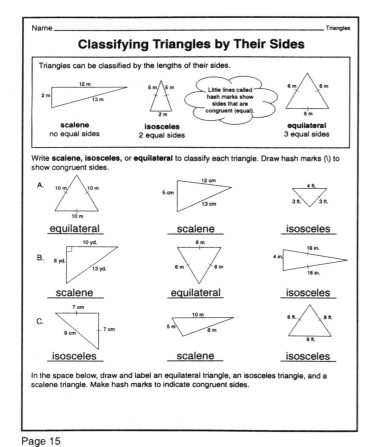

Write **scalene, isosceles,** or **equilateral** to classify each triangle. Draw hash marks (\) to show congruent sides.

A. equilateral   scalene   isosceles

B. scalene   equilateral   isosceles

C. isosceles   scalene   isosceles

In the space below, draw and label an equilateral triangle, an isosceles triangle, and a scalene triangle. Make hash marks to indicate congruent sides.

Page 15

Classifying Triangles by Their Angles

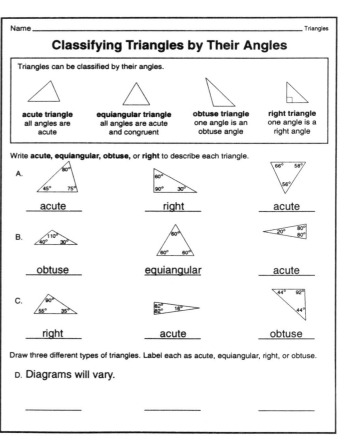

Write **acute, equiangular, obtuse,** or **right** to describe each triangle.

A. acute   right   acute

B. obtuse   equiangular   acute

C. right   acute   obtuse

Draw three different types of triangles. Label each as acute, equiangular, right, or obtuse.

D. Diagrams will vary.

Page 16

Answer Key

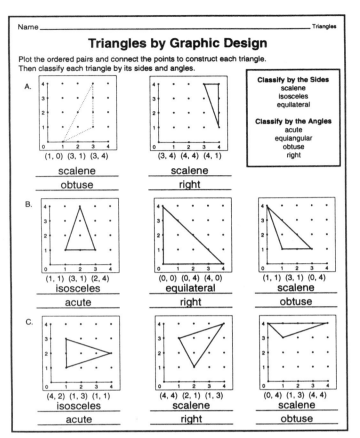

Page 17

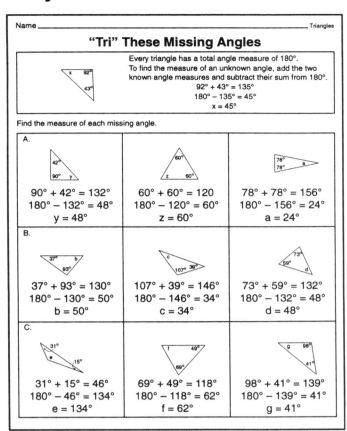

Page 18

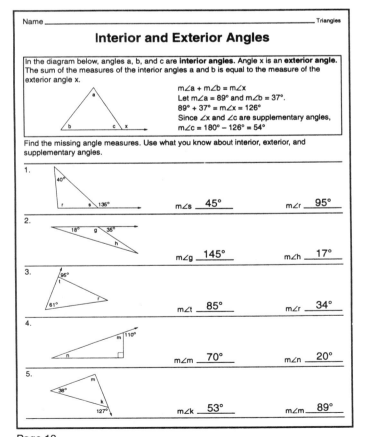

Page 19

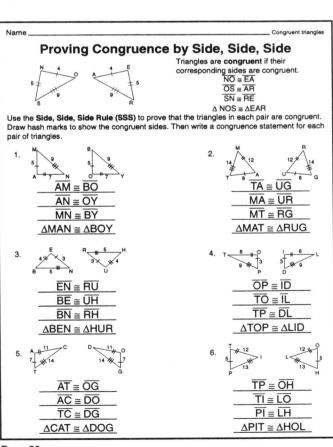

Page 20

Answer Key

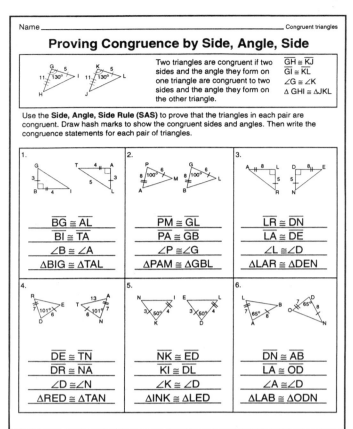

Page 21

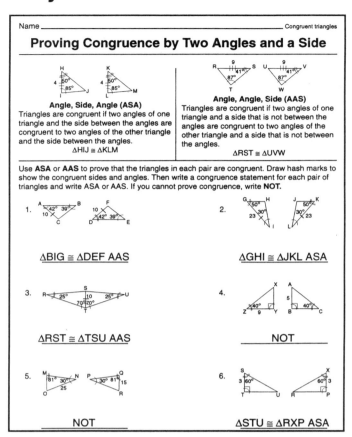

Page 22

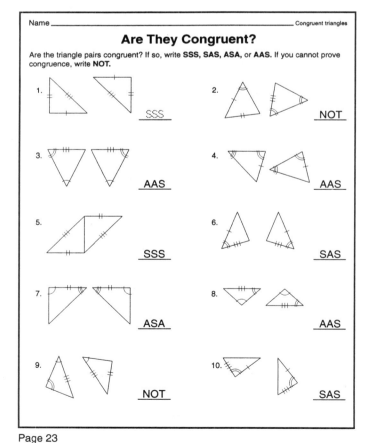

Page 23

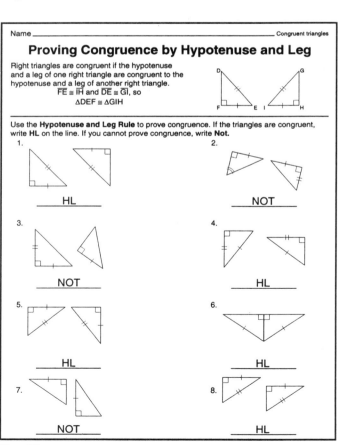

Page 24

Answer Key

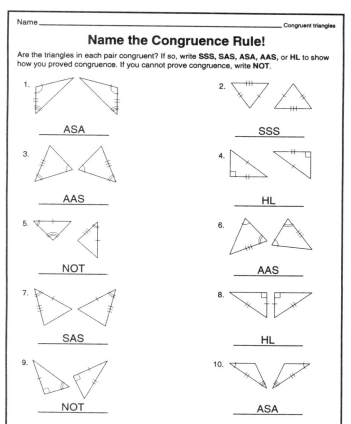

Page 25

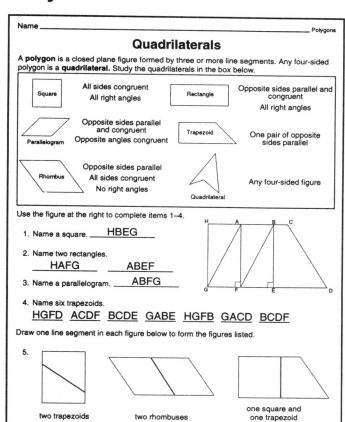

Page 26

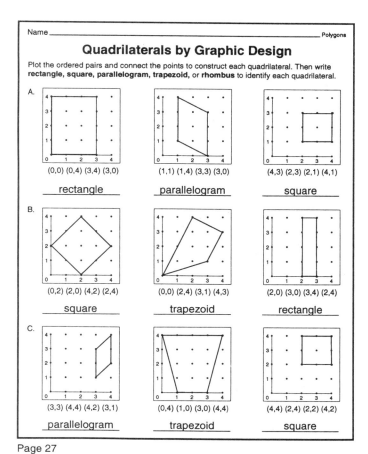

Page 27

Answer Key

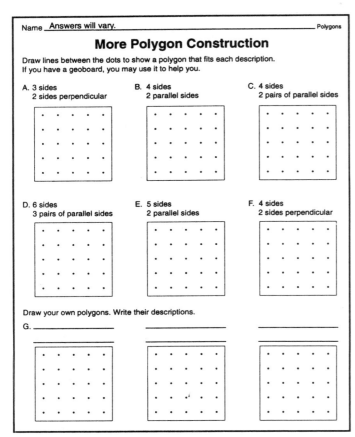

Page 29

Page 30

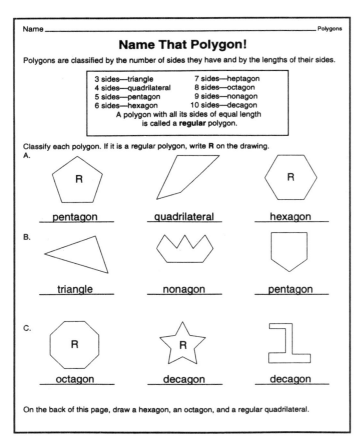

Page 31

Page 32

Answer Key

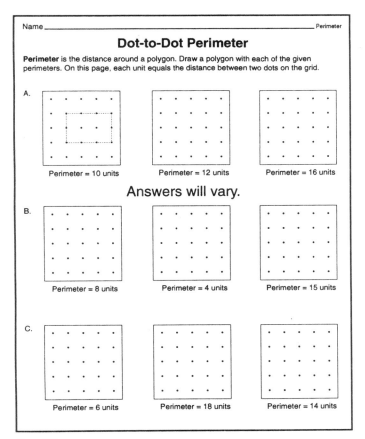

Page 33

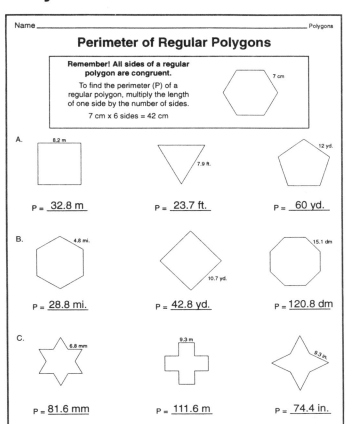

Page 34

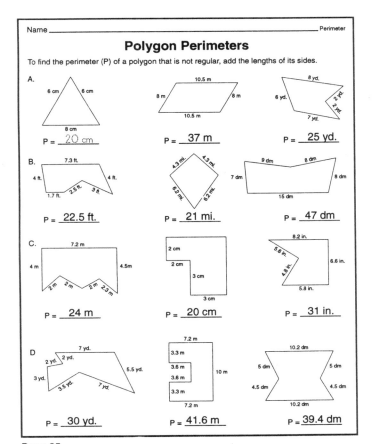

Page 35

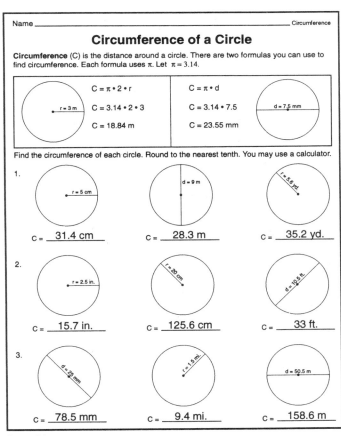

Page 36

Answer Key

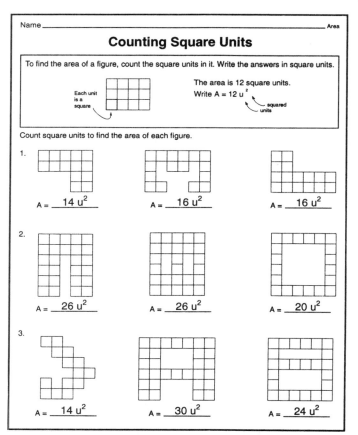

Page 37

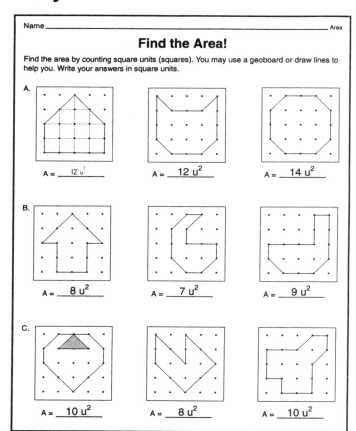

Page 38

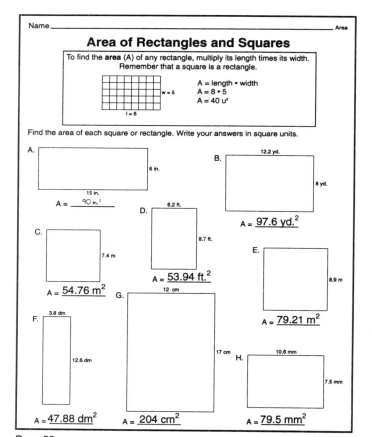

Page 39

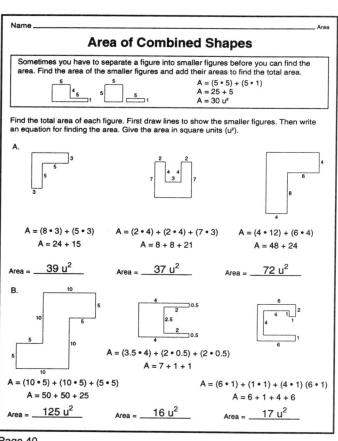

Page 40

Answer Key

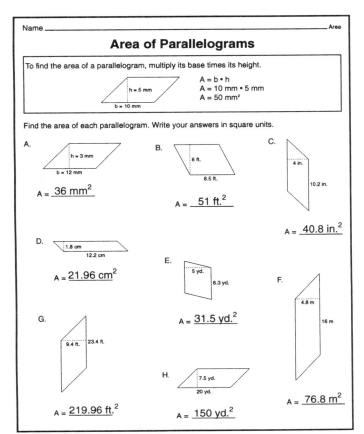

Page 41

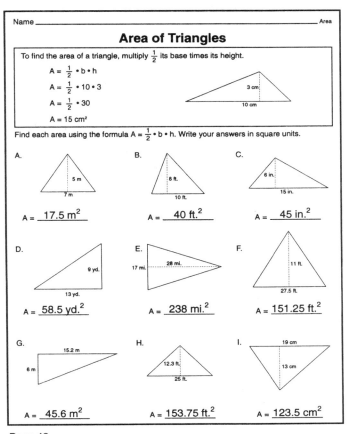

Page 42

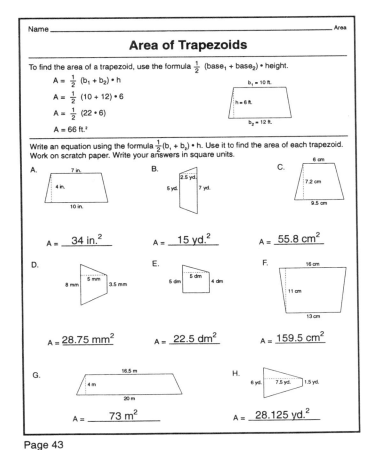

Page 43

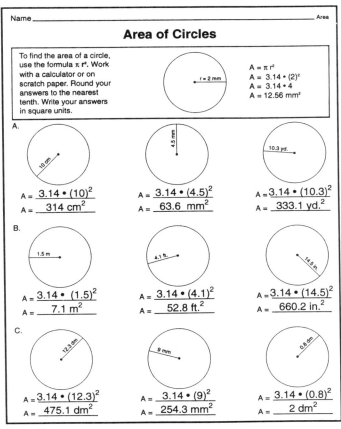

Page 44

Answer Key

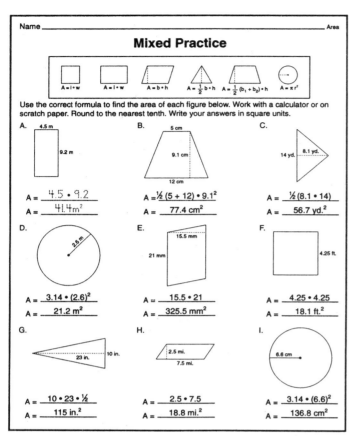

Page 45

Page 46

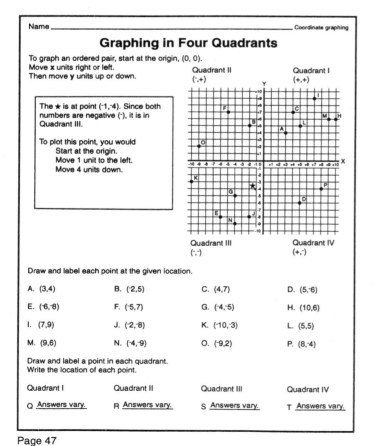

Page 47

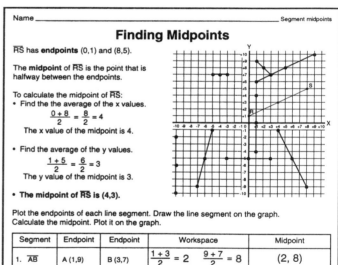

Page 48

Answer Key

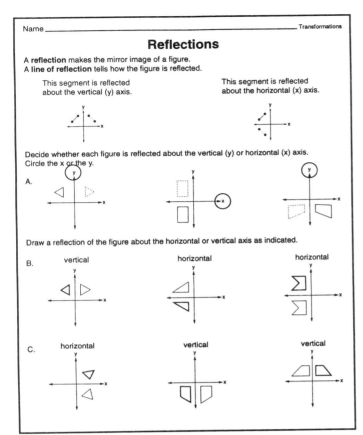

Page 49

Translations and Rotations

Is the second figure a translation or a rotation of the first figure?

A. translation — rotation — translation
B. rotation — translation — rotation
C. rotation — rotation — rotation
D. translation — translation — translation

Page 50

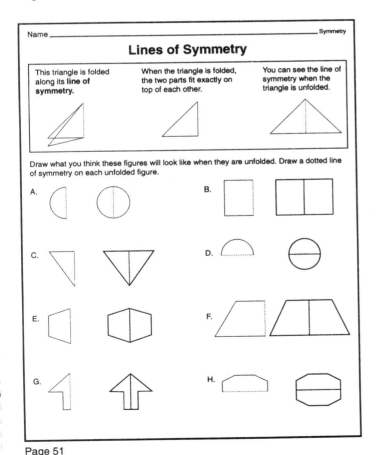

Page 51

Alphabet Symmetry

Draw and identify the lines of symmetry for the letters shown below. Write **vertical**, **horizontal**, or **both** to indicate the kind of symmetry. If there are no lines of symmetry, write **none**.

1. A — vertical; B — horizontal; C — horizontal; D — horizontal; E — horizontal
2. F — none; G — none; H — both; I — both; J — none
3. K — none; L — none; M — vertical; N — none; O — both
4. P — none; Q — none; R — none; S — none; T — vertical
5. U — vertical; V — vertical; W — vertical; X — both; Y — vertical

Page 52

Answer Key

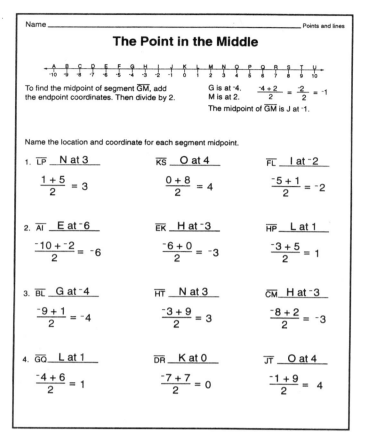

Page 53

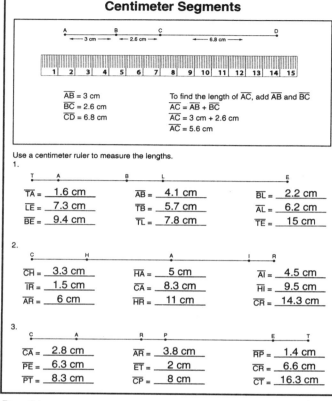

Page 54

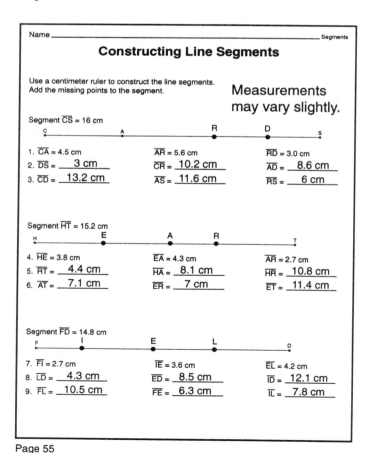

Page 55

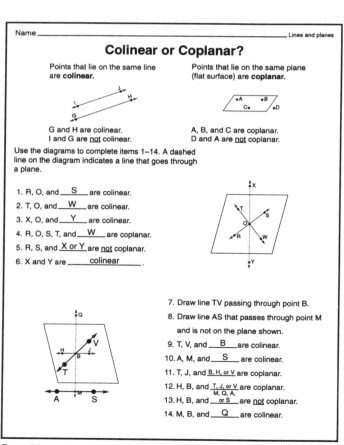

Page 56

Answer Key

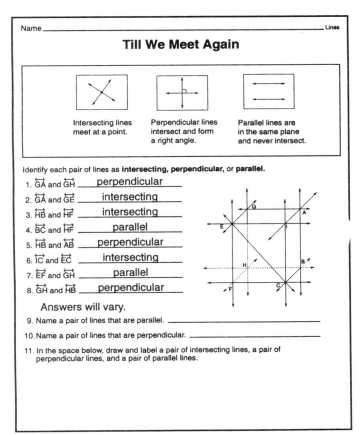

Page 57

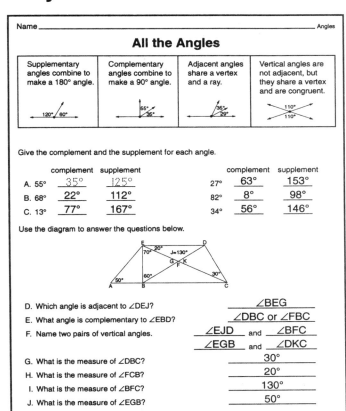

Page 58

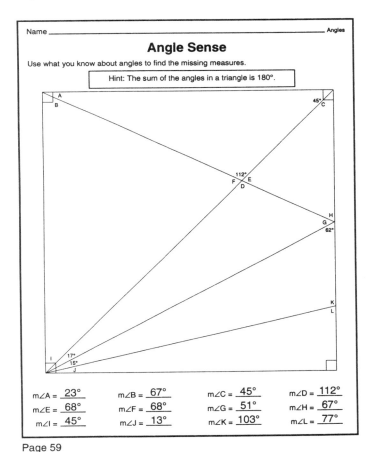

Page 59

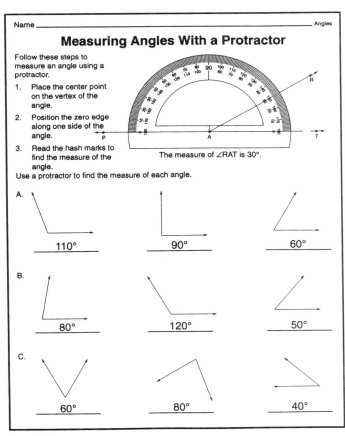

Page 60

Answer Key

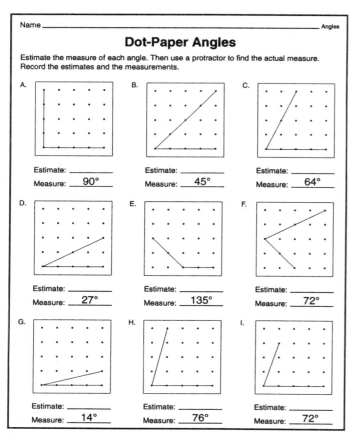

Page 61

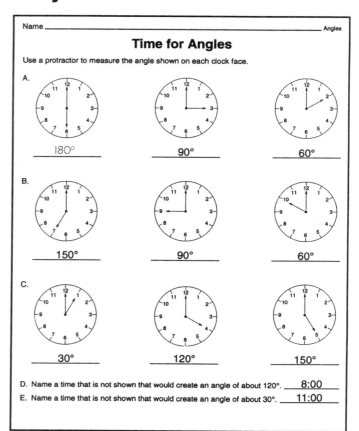

Page 62

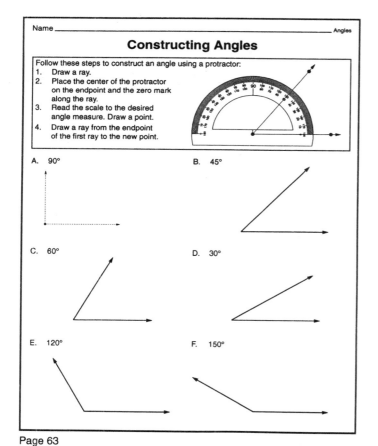

Page 63

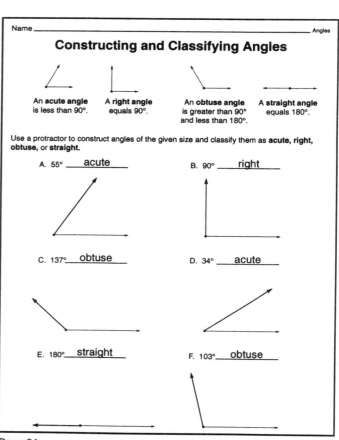

Page 64

Answer Key

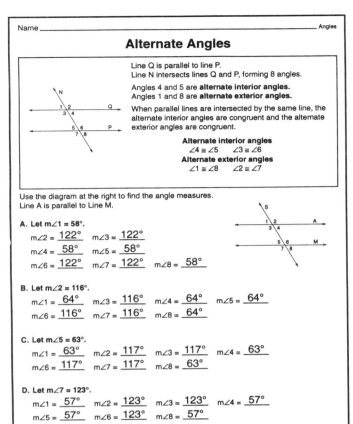

Page 65

Page 66

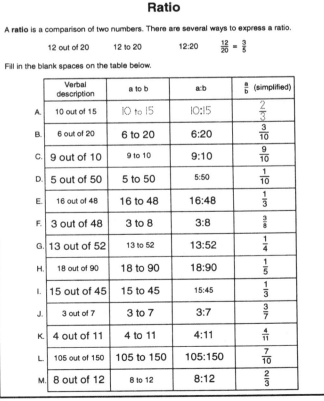

Page 67

Page 68

121

© Frank Schaffer Publications, Inc.   FS-10218 Introduction to Geometry

Answer Key

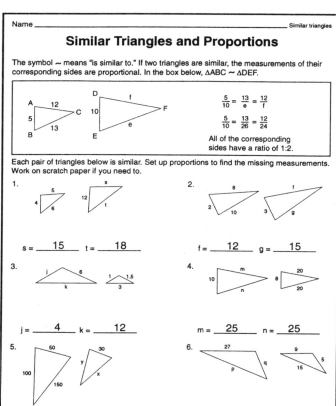

Page 69

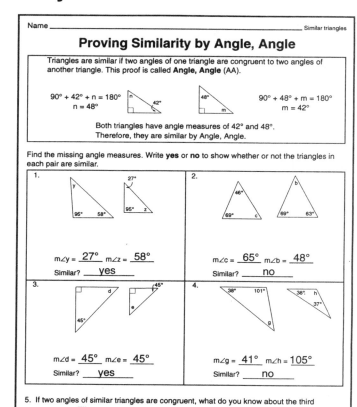

Page 70

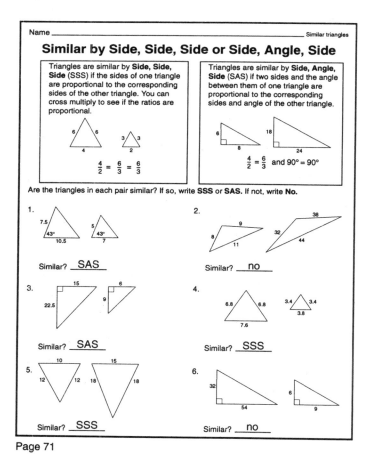

Page 71

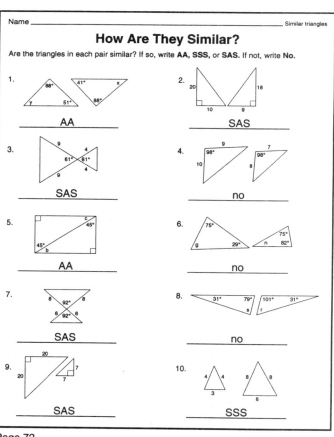

Page 72

Answer Key

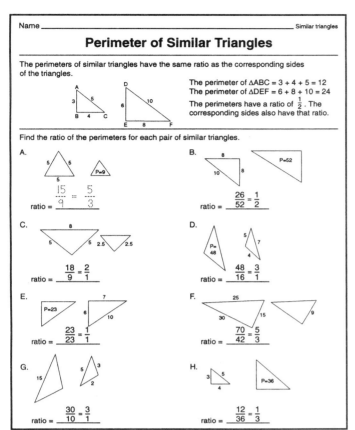

Page 73

Squared Numbers

Find the square of each number.

A.	$3^2 = 9$	$2^2 = 4$	$5^2 = 25$	$8^2 = 64$
B.	$(-4)^2 = 16$	$10^2 = 100$	$6^2 = 36$	$\left(\frac{2}{3}\right)^2 = \frac{4}{9}$
C.	$7^2 = 49$	$(-1)^2 = 1$	$\left(\frac{1}{3}\right)^2 = \frac{1}{9}$	$11^2 = 121$
D.	$(-5)^2 = 25$	$(1.1)^2 = 1.21$	$13^2 = 169$	$(-16)^2 = 256$
E.	$\left(-\frac{2}{3}\right)^2 = \frac{4}{9}$	$(0.2)^2 = 0.04$	$15^2 = 225$	$\left(\frac{1}{4}\right)^2 = \frac{1}{16}$
F.	$(0.6)^2 = 0.36$	$(1.2)^2 = 1.44$	$20^2 = 400$	$(-3)^2 = 9$
G.	$14^2 = 196$	$(-10)^2 = 100$	$(0.9)^2 = 0.81$	$19^2 = 361$
H.	$\left(-\frac{3}{4}\right)^2 = \frac{9}{16}$	$(-20)^2 = 400$	$(-0.4)^2 = 0.16$	$30^2 = 900$
I.	$\left(\frac{1}{5}\right)^2 = \frac{1}{25}$	$25^2 = 625$	$(1.3)^2 = 1.69$	$(-0.7)^2 = 0.49$
J.	$(1.4)^2 = 1.96$	$\left(-\frac{2}{5}\right)^2 = \frac{4}{25}$	$\left(-\frac{5}{6}\right)^2 = \frac{25}{36}$	$\left(\frac{6}{7}\right)^2 = \frac{36}{49}$
I.	$\left(\frac{4}{5}\right)^2 = \frac{16}{25}$	$(-1.4)^2 = 1.96$	$(-25)^2 = 625$	$\left(\frac{3}{5}\right)^2 = \frac{9}{25}$

Page 74

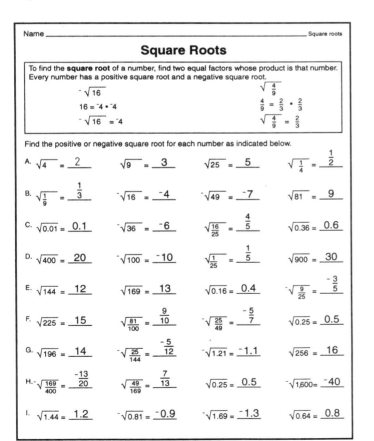

Page 75

More About Square Roots

A.	$\sqrt{2} = 1.41$	$\sqrt{10} = 3.16$	$\sqrt{15} = 3.87$	$\sqrt{3} = 1.73$
B.	$\sqrt{30} = 5.48$	$\sqrt{5} = 2.24$	$\sqrt{7} = 2.65$	$\sqrt{99} = 9.95$
C.	$\sqrt{71} = 8.43$	$\sqrt{17} = 4.12$	$\sqrt{8} = 2.83$	$\sqrt{52} = 7.21$
D.	$\sqrt{250} = 15.81$	$\sqrt{500} = 22.36$	$\sqrt{12} = 3.46$	$\sqrt{33} = 5.74$
E.	$\sqrt{75} = 8.66$	$\sqrt{150} = 12.25$	$\sqrt{21} = 4.58$	$\sqrt{40} = 6.32$
F.	$\sqrt{56} = 7.48$	$\sqrt{60} = 7.75$	$\sqrt{85} = 9.22$	$\sqrt{90} = 9.49$
G.	$\sqrt{110} = 10.49$	$\sqrt{125} = 11.18$	$\sqrt{155} = 12.45$	$\sqrt{37} = 6.08$
H.	$\sqrt{65} = 8.06$	$\sqrt{95} = 9.75$	$\sqrt{240} = 15.49$	$\sqrt{525} = 22.91$
I.	$\sqrt{80} = 8.94$	$\sqrt{600} = 24.49$	$\sqrt{825} = 28.72$	$\sqrt{130} = 11.40$
J.	$\sqrt{53} = 7.28$	$\sqrt{27} = 5.20$	$\sqrt{35} = 5.92$	$\sqrt{1,000} = 31.62$

Page 76

Answer Key

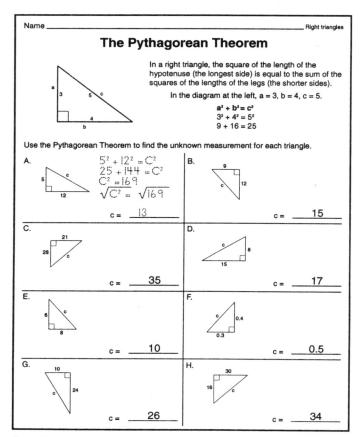

Page 77

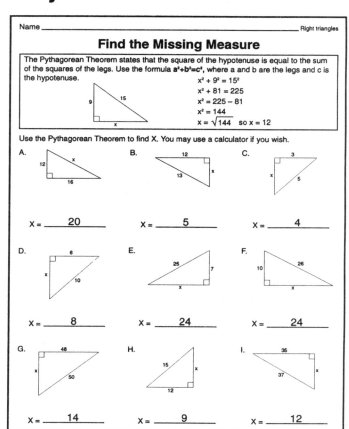

Page 78

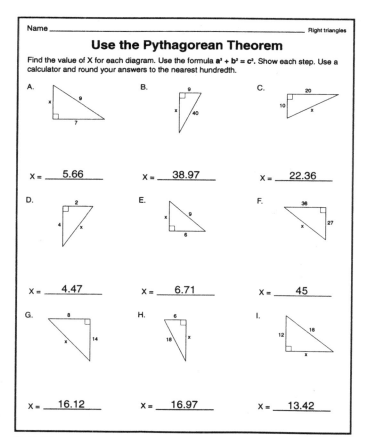

Page 79

Mixed Practice With Right Triangles

A. A right triangle has a hypotenuse of 26. The length of one leg is 15. What is the length of the other leg?
21.24

B. A triangle has a short leg with a length of 9. The other leg is twice as long. How long is the hypotenuse?
20.12

C. The legs of a right triangle measure 12 and 16. What is the length of the hypotenuse?
20

D. The hypotenuse of a right triangle is 17. One leg measures 15. What is the length of the other leg?
8

E. A triangular sail is 82 feet high. Its width is 29 feet. What is the length of the sail's hypotenuse?
86.98

F. A 25-foot ladder is leaning against a wall. It forms the hypotenuse of a right triangle. The bottom of the ladder is 6 feet from the wall. How far up the wall will the ladder reach?
24.27

G. Both legs of a right triangle measure 3. What is the length of the hypotenuse?
4.24

Page 80

Answer Key

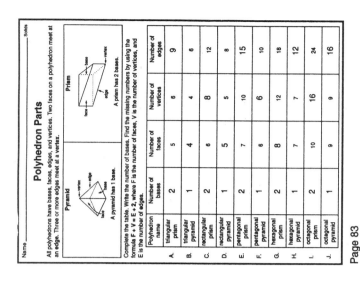

Page 81

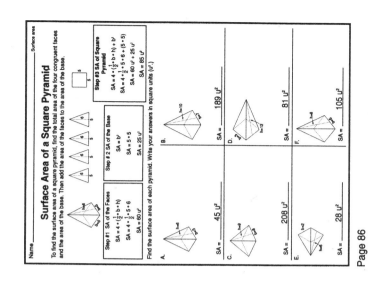

Page 83

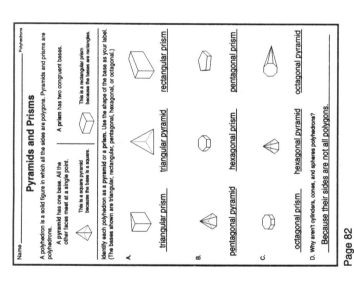

Page 82

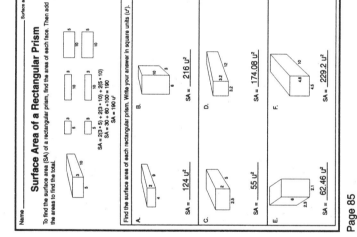

Page 85

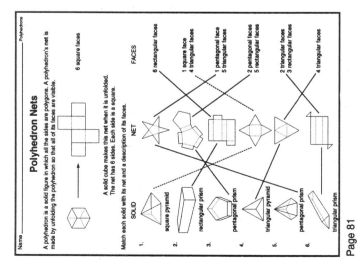

Page 84

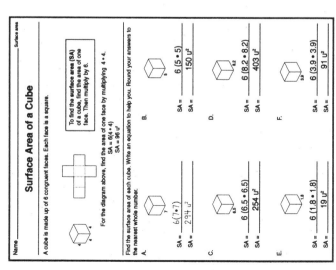

Page 86

Answer Key

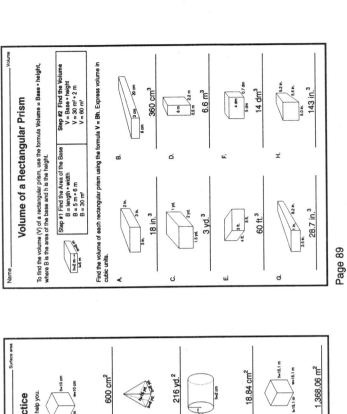
Page 89

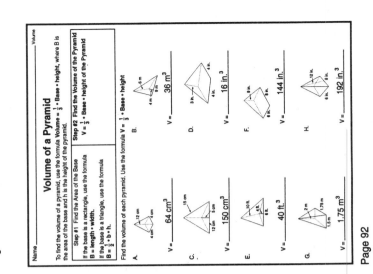
Page 92

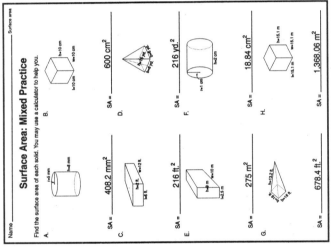
Page 88

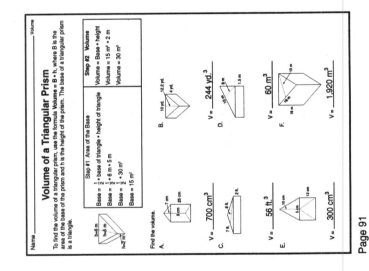
Page 91

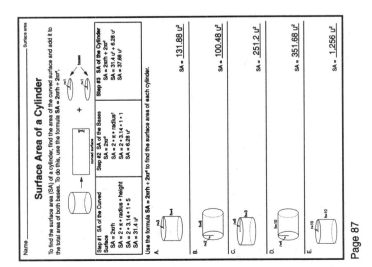
Page 87

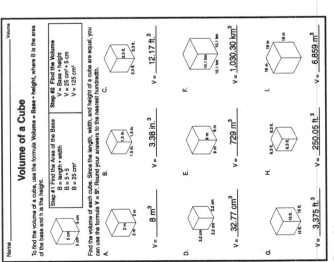
Page 90

Answer Key

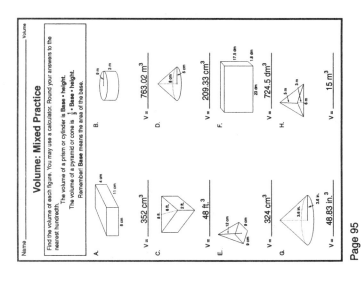

Page 95

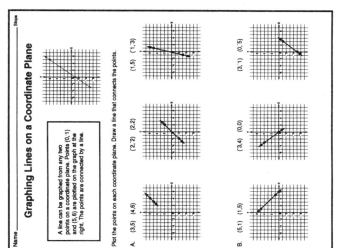

Page 98

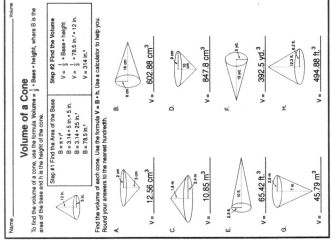

Page 94

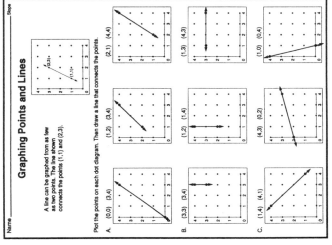

Page 97

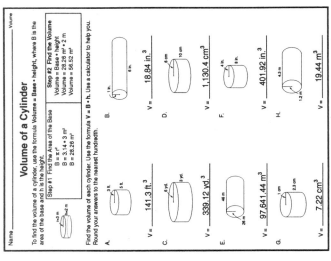

Page 93

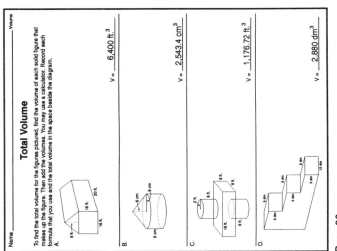

Page 96

Answer Key

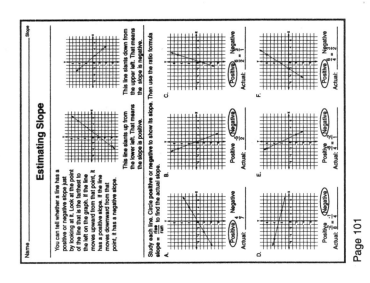

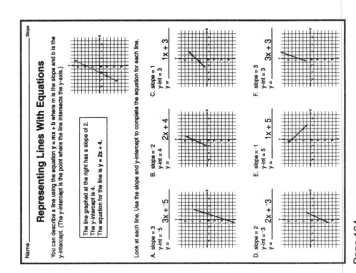

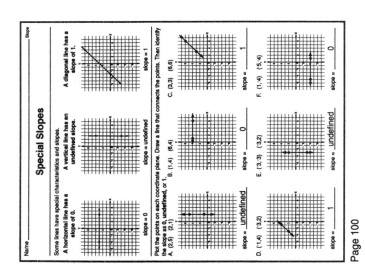

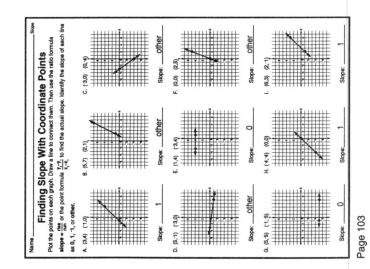

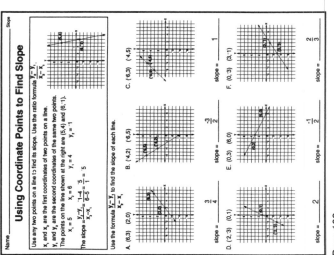